The Promise of Chemical Education: Addressing Our Students' Needs

ACS SYMPOSIUM SERIES **1193**

The Promise of Chemical Education: Addressing Our Students' Needs

Kimberlee Daus, Editor
Belmont University
Nashville, Tennessee

Rachel Rigsby, Editor
Belmont University
Nashville, Tennessee

Sponsored by the
ACS Division of Chemical Education

American Chemical Society, Washington, DC

Distributed in print by Oxford University Press

Library of Congress Cataloging-in-Publication Data

The promise of chemical education : addressing our students' needs / Kimberlee Daus, editor, Belmont University, Nashville, Tennessee, Rachel Rigsby, editor, Belmont University, Nashville, Tennessee ; sponsored by the ACS Division of Chemical Education.

pages cm. -- (ACS symposium series ; 1193)
Includes bibliographical references and index.
ISBN 978-0-8412-3092-7 (alk. paper) -- ISBN 978-0-8412-3089-7 (alk. paper) 1. Chemistry--Study and teaching. I. Daus, Kimberlee, editor. II. Rigsby, Rachel Pharris, editor. III. American Chemical Society. Division of Chemical Education.
QD40.P83 2015
540.71--dc23

2015030717

The paper used in this publication meets the minimum requirements of American National Standard for Information Sciences—Permanence of Paper for Printed Library Materials, ANSI Z39.48n1984.

Foreword

The ACS Symposium Series was first published in 1974 to provide a mechanism for publishing symposia quickly in book form. The purpose of the series is to publish timely, comprehensive books developed from the ACS sponsored symposia based on current scientific research. Occasionally, books are developed from symposia sponsored by other organizations when the topic is of keen interest to the chemistry audience.

Before agreeing to publish a book, the proposed table of contents is reviewed for appropriate and comprehensive coverage and for interest to the audience. Some papers may be excluded to better focus the book; others may be added to provide comprehensiveness. When appropriate, overview or introductory chapters are added. Drafts of chapters are peer-reviewed prior to final acceptance or rejection, and manuscripts are prepared in camera-ready format.

As a rule, only original research papers and original review papers are included in the volumes. Verbatim reproductions of previous published papers are not accepted.

ACS Books Department

Contents

Indexes

Preface

College and university faculty find themselves tasked with teaching in the face of ever-changing trends in higher education and constant shifts in the student population. Educators must balance student engagement and retention with their learning and satisfaction in a never-ending cycle of changes in technology, the economy, and the political climate. Even when certain pedagogies or classroom techniques are shown to be beneficial in one discipline, individual faculty may find it challenging to apply them in their own classrooms. This is certainly true in chemistry. Many faculty in chemistry today struggle to embrace research-based educational practices, even those coming out of our own discipline. Graduate programs in chemical education, recent reports on discipline-based education research (1), and an increase in the scholarship of teaching and learning in chemistry indicate a desire among many faculty to change—to reach students in new and exciting ways or to change curricula to better meet students' needs. Faculty are looking for things that work—techniques used by chemists, for chemists. This volume contributes to this on-going conversation.

The scholarship presented within this volume is organized in three sections. The first explores innovations found to enhance the learning of typical students as well as those who may be under-prepared. Authors describe their experiences using the flipped classroom and institutional readiness models. The second section provides examples of how technology may be utilized in the chemistry classroom—from e-textbook usage to a computational chemistry program to concrete suggestions for teaching chemistry online. The final section addresses broader issues in chemistry. One chapter demonstrates how to incorporate High-Impact Educational Practices (2) into courses for chemistry majors and nonmajors. A final chapter describes how colleges can adopt the Green Chemistry Commitment. Additionally, contextual information for pedagogical change may be found in the Introduction as well as helpful tips for adopting new approaches.

This volume is a compilation of work presented in a symposium on chemical education at the 66th Southeastern Regional Meeting of the American Chemical Society held on 17 October 2014 in Nashville, Tennessee. It represents the endeavors of faculty from small, medium, and large programs at both public and private institutions. We have included information applicable to teaching general education chemistry, freshman chemistry, and advanced chemistry majors. Each chapter provides ample resources for further research and application of the techniques presented. We hope you find the materials engaging and that you discover practical suggestions for improving your own teaching.

References

1. National Research Council. *Discipline-Based Education Research: Understanding and Improving Learning in Undergraduate Science and Education*; National Academies Press: Washington, DC, 2012.
2. Kuh, G. D., Schneider, C. G. *High-Impact Educational Practices: What They Are, Who Has Access to Them, and Why They Matter*; Association of American Colleges and Universities: Washington, DC, 2008.

Kimberlee Daus

kim.daus@belmont.edu (e-mail)
Department of Chemistry
Belmont University
1900 Belmont Blvd.
Nashville, Tennessee 37212, United States

Rachel Rigsby

rachel.rigsby@belmont.edu (e-mail)
Department of Chemistry
Belmont University
1900 Belmont Blvd.
Nashville, Tennessee 37212, United States

Chapter 1

An Introduction to Educational Promises: Challenges and Strategies

Kimberlee Daus* and Rachel Rigsby*

**Department of Chemistry, Belmont University,
Nashville, Tennessee 37212, United States
*E-mail: kim.daus@belmont.edu; rachel.rigsby@belmont.edu**

Adopting an alternative pedagogy in your classroom can be a daunting task—options seem endless, and barriers to change are high. Here we frame evidence-based strategies presented in this volume around identified high-impact educational practices. Additionally, we provide tools to guide the decision-making process for faculty looking to make changes in their classrooms.

Introduction

Congratulations! Whether you've decided to make changes to your course or program due to a personal desire to increase learning in your classroom or due to institutional programmatic changes, we welcome you to the engaging conversation surrounding teaching and learning chemistry. This chapter is designed to help you on your journey. First, we will consider one call to action that broadly impacts teaching and learning in any program. Second, we will look at challenges you may face in adopting new pedagogy or strategies to address the challenges. Finally, we will help you match your current needs with chapters in this volume using the criteria of course placement and multiple high-impact educational practices.

A Call for Reform

Over the years there have been many calls for educational reform. A recent challenge, issued by the American Association of Colleges & Universities, invites teachers and administration to rethink curriculum in terms of pedagogical practices. In *High-Impact Educational Practices: What They Are, Who Has Access to Them, and Why They Matter,* George Kuh identifies ten pedagogical approaches that have proven beneficial for many college students (*1*). Several of these approaches, which Kuh termed "high-impact practices," are well known to chemists; these include undergraduate research, collaborative projects and assignments, internships, and capstone courses and projects. Other practices may be utilized by some chemists in their courses or curriculum (learning communities, writing-intensive courses, service- or community-based learning, and common intellectual experiences). Additional practices, including first-year seminars and diversity/global learning, may be adopted as part of a chemistry curriculum and/or may be taught by chemistry faculty. These practices, when incorporated into the chemistry curriculum and courses, can result in higher levels of student engagement and critical thinking skills. Such practices can also help students make better connections between content areas as well as provide them with applicable experiential learning.

Challenges and Barriers

Decades of research funded by organizations such as the National Science Foundation have proven the effectiveness of many teaching strategies, with clear evidence supporting increases in student learning and engagement. In addition to the practices identified by Kuh, specific innovations in science pedagogy are abundant. Even the briefest literature review offers thousands of results touting the advantages of problem-based learning, active learning, process-oriented guided inquiry learning (POGIL), and even a recent article on 'innovation pedagogy' (*2*). However, faculty adoption of new strategies is slow (*3*). This problem is not unique to chemistry—reports describe the resistance of medical schools to adopt problem-based learning, which began in 1968 but has taken years to achieve wide-spread acceptance despite solid evidence of its effectiveness (*4*). One case study in chemical engineering posits that convincing research doesn't result in faculty adoption of new pedagogy (*5*). Why do faculty resist change? Anecdotal evidence (possibly your experience as well) would suggest a lack of time or resources as primary barriers. Academicians are busy, balancing teaching responsibilities with a myriad of activities including research and publication expectations, student advising, and institutional service. Additionally, they are notoriously autonomous, with strong thoughts on what should go on in their classrooms. Fundamentally, faculty often tend to teach how they were taught, which for many in the sciences was the standard 'sage on the stage' lecture method. One report suggests that a primary barrier producing this resistance is that STEM change strategies are primarily based on a development and dissemination change model (*6*). This is a 'top-down' approach to STEM education—faculty are handed, in the form of scholarly publications or suggestions from administrators, strategies 'proven' to improve student learning. However, faculty are often not armed with practical suggestions for implementing changes in the context of their individual curriculum or classroom or have the necessary support structure to aid them in making changes. Furthermore, faculty in the tenure pipeline may perceive the risk of change as being too great (3). The result is, in spite of funding to develop and disseminate new pedagogies, the current model of prescribing change and expecting implementation doesn't produce the desired change.

So, how then do we implement change? Most faculty have a desire to improve their teaching. However, many faculty feel overwhelmed at the challenges associated with change. You may find yourself in this situation—looking for ways to invigorate your teaching but not sure where to begin. Sometimes looking for help from colleagues is a great place to start. A recent five-year project funded by the National Science Foundation aims to distribute new pedagogies to pre-college teachers through networks, where teachers demonstrate techniques for each other and encourage everyone in their network to use best practices in their classrooms (7). It seems that seeking assistance from others in the trenches of teaching can be an effective agent of change, bringing new pedagogies to more classrooms in a timely manner.

Now that you've become interested in pedagogical change, you may be asking yourself, "What is the 'best' strategy to use in my classroom?" According to Ken Bain, author of "What the Best College Teachers Do," it takes more than just adopting a pedagogy to create a successful learning environment (8). In addition to knowing their subject matter well, the best teachers approach their classes as "serious intellectual endeavors (8)." They have high but realistic expectations for their students and are able to create a strong, appropriate rapport with students. Their assignments focus on relevant, timely questions in their disciplines and the teachers are able to share their own sense of wonder and curiosity (in their fields). The best teachers are reflective and honest with themselves and, if an endeavor fails, look for the source of failure and do not blame their students. In other words, it's not the specific pedagogy that creates the optimal learning environment; what is important is how effectively the teacher uses that pedagogy to reach students.

Practical Advice

Finding that it doesn't matter what you choose may take the pressure off! On the other hand, if there isn't one 'best' option, what should you choose, given the plethora of pedagogical choices available? The following suggestions may be helpful as you consider how to implement change in your classroom:

- Be true to yourself. Choose a pedagogy that you are passionate about and that you would be comfortable implementing. Once you've been successful in one particular pedagogy you may be confident enough to branch out and try others.
- Start small. Changing one assignment, trying a new lab format, or implementing one week of flipped instruction in one class allows you to trial a new pedagogy to see what works for you.

- Communicate with your students. Talk with them about why you are trying this new teaching endeavor, what you hope they gain from it, and evidence that supports your trial. Also, clearly identify learning objectives that this new technique will address.
- Ask for student feedback. Ask students what worked, what didn't work, and for any suggestions for improvement.
- Be reflective. In addition to taking notes on what worked and what didn't, it is good practice to also personally reflect on the experience. Were you comfortable implementing the pedagogy? What did you like and/or dislike about the experience? Were the students more engaged as a result of the pedagogy? Did deeper learning result? How do you know?

How To Use the Book

We hope this volume provides you with resources to begin or further your journey of change in the classroom. For a quick view of the book and its topics, Table 1 organizes the strategies presented in this work with their intended audience of non-science majors, freshman or sophomore chemistry majors, or upperclassmen. Additionally, the table identifies strategies applicable across an entire program or curriculum. The table further aligns each option with one or more related High-Impact Practices (HIPs) (1). Brief descriptions of the HIPs (9) and connections to chapters in this volume immediately follow the table.

First-Year Seminars and Experiences

Many programs are beginning to use first-year courses to bring small groups of students together with faculty or staff on a regular basis. In chemistry, they may be used to introduce faculty research or help students adjust to college life. Information in Chapter 2 (Teaching to the Edges) could easily be used in a first-year experience. Additionally, Chapter 4 describes the use of e-textbooks in a first year chemistry class.

Common Intellectual Experiences

The idea of a "core" curriculum has evolved into a variety of forms, such as a set of required common courses for students moving through a program. This could also be thought of as a common theme or pedagogy used by faculty at various points within a program. In the context of a chemistry curriculum, online discussion boards (Chapter 5) and a specialized software tool for students (Chapter 7) could provide the educational advantages of a common experience.

Table 1. Alignment of pedagogies in this work with their targeted student group. HIPs: 1 = First-Year Seminars and Experiences; 2 = Common Intellectual Experiences; 3 = Learning Communities; 4 = Writing Intensive Courses; 5 = Collaborative Assignments and Projects; 6 = Undergraduate Research; 7 = Diversity/Global Learning; 8 = Service Learning/Community-Based Learning; 9 = Internships; 10 = Capstone Courses and Projects

	Nonmajors	Majors – Freshman	Majors – Sophomores	Majors – Upperclassman	Programmatic	HIP
Teaching Chemistry to the edges		✓				1
Flipped Classroom - organic			✓			5
E-textbooks	✓	✓	✓	✓	✓	1
Discussion board	✓	✓	✓	✓		2
Research methods course				✓		6, 10
PSI4 Education			✓	✓		2, 6
Experiential Learning	✓	✓	✓	✓	✓	3, 8, 9
Green Chemistry	✓	✓	✓	✓	✓	7, 10

Learning Communities

Learning Communities are designed to encourage integration of learning across different courses. Students take two or more linked courses as a group and work closely with one another and with their professors. Learning Communities can explore a common topic and/or common readings through the lenses of different disciplines. Information on general education and major-specific chemistry learning community courses (Chapter 8) could be helpful if you are looking to incorporate this pedagogy into your curriculum.

Writing-Intensive Courses

These courses emphasize writing across the curriculum. Some programs may have writing-intensive courses in the form of classes designed to teach students scientific writing through production of lab reports or undergraduate research papers. While not addressed here, options specific to science can be found in the literature (*10, 11*).

Collaborative Assignments and Projects

Collaborative learning helps students learn to solve problems as part of a team. This can occur in a variety of ways, from course study groups to team assignments, and even undergraduate research. Using the flipped classroom approach (Chapter 3) can deepen collaborative relationships as students work together in the classroom.

Undergraduate Research

Research is a component of many programs in the sciences. Many programs are funded by the National Science Foundation (*12*) and supported by the American Chemical Society's undergraduate research symposia. Typical chemistry research courses could be modified using techniques outlined in Chapters 6 (Methods Course) and 7 (PSI4).

Diversity/Global Learning

Many colleges approach diversity and global learning from the perspective of teaching students about other cultures and worldviews. While cultural viewpoints are often neglected in a typical chemistry curriculum, the adoption of the Green Chemistry perspective (Chapter 9) certainly applies here.

Service Learning, Community-Based Learning

The goal of this pedagogy is to provide students with direct, real-world applications of their discipline where they experience deeper learning of subject matter through service to their community. Many faculty may dip into this arena in a co-curricular way through ACS outreach programs through their student groups. A key component to using this in the classroom is to add learning goals and student reflections on the experience. The chapter on experiential learning (Chapter 8) can provide ideas for learning and applying chemistry in your community.

Internships

Internships are another increasingly common form of experiential learning. Here students are supervised as they experience their discipline in a professional environment. Internships are often used in addition to undergraduate research to give students hands-on experience in chemistry. We do not directly address the use of internships in chemistry in this work.

Capstone Courses and Projects

Whether they're called "senior capstones" or some other name, these culminating experiences require students nearing the end of their college years to create a project of some sort that integrates and applies what they've learned. The project might be a research paper, a performance, a portfolio of "best work," or an exhibit of artwork. Capstones are offered both in departmental programs and, increasingly, in general education as well. Examples of capstone courses/projects may be found in several chapters (Chapters 6 and 9).

Conclusion

So, is changing your classroom worth the cost? It will take a lot of work, but we think it is. We hope you find the information presented here and in the chapters that follow helpful and inspiring. We wish you the best of luck in your teaching endeavors!

References

1. Kuh, G. D.; Schneider, C. G. *High-Impact Educational Practices: What They Are, Who Has Access to Them, and Why They Matter*; Association of American Colleges and Universities: Washington, DC, 2008.
2. Kettunen, J. *Creative Education 2.1* **2011**, 56–62.
3. Tagg, J. *Change* **2012**, *44*, 6–15.
4. Jippes, M.; Majoor, G. D. *Medical Education* **2008**, *42*, 279–285.
5. Golter, P. B.; Thiessen, D. B.; Van Wei, B. J.; Brown, G. R. *J of STEM Education* **2012**, *13*, 52–59.
6. *Barriers and Promises in STEM Reform*; Commissioned Paper for National Academies of Science Workshop on Linking Evidence and Promising Practices in STEM Undergraduate Education, Washington, DC, October 13–14, 2008.
7. EnLiST: Entrepreneurial Leadership in STEM Teaching and learning. http://enlist.illinois.edu/ (accessed June 30, 2015)

8. Bain, K. *What the Best College Teachers Do*; Harvard University Press: Cambridge, MA, 2004; pp 15–20.

9. High-Impact Educational Practices. http://www.ou.edu/ae/highimpact.html (accessed July 1, 2015).

10. Whalen, R. J.; Zare, R. N. *J. Chem. Educ.* **2003**, *80*, 904–906.

11. Writing-Intensive Courses. http://www.wpi.edu/academics/cxc/intensive-courses.html (accessed July 2, 2015).

12. National Science Foundation. http://www.nsf.gov/crssprgm/reu/reu_search.jsp (accessed July 2, 2015).

Classroom Innovations

Chapter 2

Teaching College Chemistry to the Edges Rather Than to the Average

Implications for Less Experienced Science Students

Dorian A. Canelas*

Department of Chemistry, Duke University,
Durham, North Carolina 27708-0346, United States
*E-mail: dorian.canelas@duke.edu

Students in the lowest quartile of their matriculating class in terms of the math and science experiences usually struggle in their first chemistry course in college. Although math and science experience closely tracks with family income and access to advanced placement and honors high school courses rather than with individual intelligence, attrition from science coursework during the first year of study has become the norm for this group. Access to college continues to expand, but institutions of higher learning have been slow to adapt to the increasingly diverse backgrounds of their student bodies. The challenge presented by the breadth of learners' backgrounds constitutes an issue of poor institutional readiness rather than a problem arising from any characteristics of the students themselves. This viewpoint puts the duty to provide legitimate pathways for student success squarely on the shoulders of faculty who develop curricula, teach classes, and lead programs. With this in mind, faculty in the Duke University Department of Chemistry began systematic assessment of undergraduate student outcomes in gateway chemistry courses. Results from this work spurred major curricular and institutional programming changes to address the wide range of learner backgrounds in chemistry and scientific problem solving. Better matching of courses with the learners' existing skills and implementation of known high impact practices such as collaborative in-class work and new learning communities were keys in this process. A preliminary assessment of these changes, including demographic impact, will be discussed.

Societal and income barriers to post-secondary education are dissolving. This challenges institutions to implement programs and readiness models that address the wide variability of students' experiences, skills, and backgrounds. Despite the logistical hurdles, time commitment, and expense in the implementation of successful programs to this end, gains have been made in increasing the diversity of the learners earning degrees in science, technology, engineering, and mathematics (STEM) fields in the past 50 years (*1*). But the work remains unfinished: high course attrition rates of students from all backgrounds continue to plague STEM programs at colleges and universities nationwide (*2, 3*). Substantial additional progress is needed to achieve racial and gender parity in retention through the pipeline (*4*), so national efforts aimed at increasing diversity in STEM in the first two years of college continue (*5*).

Underrepresented minorities (URMs), defined here as African Americans, Latinos, and Native Americans, represent nearly 30 percent of the US population and almost 40 percent of the nation's K–12 enrollment (*6, 7*). However, this demographic group earns less than 20 percent of the bachelor's degrees and less than 10 percent of the doctorates in STEM (*6, 7*). This gap in the composition of the STEM professional ranks is *not* due to lack of interest in science on the part of high school students and matriculating college students. As part of a quantitative comparison of a diverse group of learners at key transition points, Garrison notes that "Among college freshmen, race-ethnic differences in plans for a science or engineering major are very small and have little impact on the ultimate level of underrepresentation (*8*). . ." Moreover, this segment of the population is growing: 2011 was the first year in which domestic births of white babies were not the majority according to data released by the United States census bureau (*9*). In fact, people in groups currently described as URMs will actually be the majority of the population within just a few decades. So, efforts to open real access to advanced training in the STEM disciplines need to continue not only because of the tenant of equal opportunity upon which this country was built, but also because of the practical need for the United States to remain a leader in science, innovation, and healthcare in the global economy.

Why is chemistry, just one of many STEM disciplines, especially important in this discussion? One reason is that chemistry is often described as the "central science," with textbooks and even an ACS journal noting this distinction in their titles. A basic understanding of chemistry is needed for advanced study in many other disciplines, but it is not especially emphasized in our K-12 education system. Instead, most students enter college with merely a single year of high school chemistry coursework under their belts. When coursework is mentioned, college chemistry courses are much more frequently cited by students than any other courses in interviews about abandoning pre-medical aspirations (*10, 11*). Barr concludes that chemistry is the *key* gateway course sequence that discourages students from continuing to pursue science and health-related degrees (*10, 11*). Barr also notes that, in his studies of undergraduates at Stanford and the University of California at Berkeley, this phenomenon severely and disproportionately impacted female students and URM students (*10, 11*). This point is of particular importance at Duke University because of the way our curriculum is currently structured: credit for college-level general chemistry is a pre-requisite for the gateway biology courses. Because of this requirement, first-term freshman interested in science are taking either chemistry or physics as their first college science course (and the vast majority, >90%, take chemistry first.)

Despite the documented history of some discouraging effects of chemistry coursework on science major selection and pre-health study, best practices for the college academic and co-curricular experiences that lead to retention of all students, and URM students in particular, in STEM undergraduate education have been well documented (*12–15*). Chang found that "studying frequently with others, participating in undergraduate research, and involvement in academic clubs or organizations" were student behaviors that correlated with increased persistence in STEM coursework (*12*). Herein, a practioner's view of teaching, departmental curriculum reform, and the creation of student communities of scholarship is presented. In addition to a discussion of the early outcomes from assessment of chemistry curriculum reform, the implementation of two key high impact practices (*15*), *collaborative assignments and projects* and *learning communities*, will be discussed in the context of their correlation with student success and retention in chemistry.

Theoretical Framework and Perspective

From the extreme viewpoints, the relatively high attrition rates from STEM coursework pathways can be viewed as either a student-centered issue or an issue of institutional readiness. This critical choice of perspective governs how curricular and student support programming choices are made by STEM department faculty, university teaching and learning centers, and the institutional administrators who fund these choices. Each of these perspectives is explored in more detail below.

Student-Centered Models for Explaining High Attrition

Researchers invoking student-centered models have explored various angles that contribute to students staying in science or leaving science. One of the most prominent faculty viewpoints is the *student deficit model* of thinking. Proponents of this model subscribe to the perspective that many or most students who leave STEM fields, even after expressing strong interest immediately prior to matriculation, do so because they are either woefully academically underprepared (*16–18*) or because they expect high grades yet are unwilling or unable to tackle the heavy academic workload (*19, 20*). This model conveniently allows college educators to absolve themselves of any responsibility for high attrition or failure rates in their own classes and evoke an external locus of control. After all, in this model for explaining attrition, all fault for failure to thrive lies with the either the students (in some cases entire demographic groups) or the secondary school system from whence they graduated. Some professors quite enthusiastically cling to this model, feeling that only "above average" or the "strongest" (in their estimation) students should "survive" in their courses. Koebler describes this attitude: "Many veteran STEM professors believe science should be hard, and the course work isn't something every student can do. For them, difficult freshman-year classes separate the cream of the crop (21)." Infused in this viewpoint are three co-existing central premises: 1) that raw talent and intrinsic ability are the most important characteristics for excelling in fields such as science, 2) that ability is fixed and innate in each individual, and 3) that individuals with certain demographic characteristics (e.g. Caucasian male or Asian male) (22) are most likely to possess this needed talent. Although many academics would flat out deny having the implicit bias apparent in the third point, a recent nationwide survey of academics confirmed the pervasiveness of this philosophical vantage point, termed "field-specific ability beliefs", in college and university faculty (22).

Other student-centered models for attrition from STEM focus on student attitudes about STEM fields. Student perceptions about how each individual or his or her group fits into a specific field are examined. Numerous studies have explored the impact of stereotype threat–"the anxiety of being judged based on a negative group stereotype (23)"–and more recently "expectations of brilliance" commonly associated with science, mathematics, and some other academic fields, such as philosophy, economics, and music composition (22).

In a variation of the student-deficit theory, students who are in the bottom quartile of their class in terms of their pre-college standardized test scores are said in some circles to be "mismatched." Decades ago, proponents of the "mismatch hypothesis" asserted that such students experience relative deprivation that causes them to change career paths (24). The message was that these students should have chosen to attend less selective institutions in which they would have been a "big fish in a small pond (24)." More recently, Smyth and McArdle asserted that "At the individual level, offering a relatively educationally disadvantaged applicant the chance to benefit and graduate from a more selective institution may put at increased risk his or her goal of a career in science (25)." Espinosa confirmed these results in the specific case of women of color in undergraduate STEM majors: "negative indicators of persistence include attending a highly selective institution (26)." On the other hand, in a much larger study, Chang and co-workers showed that the situation is quite complicated, and that the mismatch hypothesis and anticipatory socialization theory counter each other (27). They recommend that "research universities should take a much harder look at why those students who should otherwise complete a science major are not doing so on their campuses (27)."

Institutional Readiness Models for High Attrition

The counterpoint to this furor over student deficits, stereotype threat, and mismatch hypothesis is the concept of needed improvements in institutional readiness. Proponents of the institutional readiness model (28) assert that our academic institutions still erroneously function via a curricular framework designed for a demographically much more homogeneous student body of decades past. Indeed, it was probably quite reasonable to put all students in a single introductory chemistry or biology or French class at a highly selective college when all students had remarkably similar high school experiences at a narrow subset of the nation's strongest high schools. But, teaching to the "average" only works when deviations from the average are minor. When multiple dimensions are considered for a diverse group, the likelihood of any individual being near or above average in all dimensions becomes extremely small (*29*). In the same way, it was logical to design U.S. military aircraft cockpit dimensions to male height and weight specifications when all pilots were men. Prior to 1993, these dimensions accommodated the sitting height of 90% of white men but only 30% of women (*30*). The integration of women into pilot ranks was a disruptive change, challenging the status quo and forcing adaptation (*30*). Mandated changes to military cockpit design parameters benefited not just women, but also men of smaller stature. Because of this, it disproportionately impacted men in certain racial and ethnic groups. Even before the changes were mandated, pragmatists in the military argued that such changes would eventually be necessary regardless of the combat status of women given the changes in the ethnic and racial composition of the nation's population (*31*).

Similarly, increased access to higher education means that college-bound students hail from a much wider variety of backgrounds than ever before. Indeed, the diversity of learners in college classrooms has expanded dramatically in the past fifty years. The institutional readiness viewpoint rejects the ideas of the student deficit model, mismatch hypothesis, or any simple definition of an average or below average student, recognizing that each student brings a unique set of strengths and weaknesses to his or her course work (*32*). This mode of thinking, sometimes framed as jagged learner profiles, allows for the fact that the student who has the most potential to be a truly exceptional scientist might enter college with less prior exposure to science learning than his or her peers. As an illustration, a particular student might enter college above average in creativity, work ethic, self-confidence, intrinsic motivation, and quantitative reasoning but below average in spatial reasoning, analytical reading, writing, self-regulation, and interpersonal skills compared to his or her peers. Students can be clustered into more narrow learning profile groups based upon similarities in approaches to retaining information and problem solving (*33, 34*), and changes to these learning profiles can be tracked over time (*34*). Indeed, all students have relative strengths and weaknesses in areas important to academic learning and ultimately in becoming a successful scientist. The institutional readiness model puts responsibility to address these differences and develop a more highly functional educational system firmly on the shoulders of university faculty and administrators. In the ideal, more highly functional institutional system, motivated students from all backgrounds should have an authentic opportunity to succeed in STEM gateway courses and majors. This perspective is the one that motivates us, as college and university educators, to make changes to the design of our classes and programs in ways that allow the widest possible sample of our more heterogeneous student body attending college in the 21st century to thrive.

Using this theoretical framework and perspective, this assessment focused on students with math SAT scores in the lowest quartile of their matriculating class. It aimed to determine the baseline academic situation for these students in chemistry courses given exposition-style classrooms and the curriculum prior to 2009. From there, it aimed to explore how students' pre-matriculation academic backgrounds related to their grades and persistence in the initial gateway chemistry courses after curriculum changes and to determine whether or not these relationships varied as a function of gender and URM status.

Methods and Descriptions of Courses

Demographic and chemistry course outcome data were mined by personnel of Duke University's Trinity College Office of Assessment. Similar sets of data were mined for students in the control group (Group 1), who matriculated in the 2006 or 2007 academic years, and students experiencing the new curriculum and programming (Group 2), who matriculated in the 2009 or 2010 academic years. Both groups were comprised only of students who enrolled in chemistry and who were in the lowest quartile of their matriculating class in terms of reported SAT math score (in some cases the SAT score was not reported, so when possible the ACT math score was converted using a percentile-based concordance table) (*35*). None of the students in either group had a reported score of 5 on the Advanced Placement (AP) chemistry test. Students with a reported AP chemistry score of 4 ($N = 4$) were excluded from the study prior to analysis because those students were qualified to start in a higher level chemistry class (*Honors Chemistry*) according to university placement guidelines.

All students in Group 1 began their chemistry study with Chem21L, first semester *General Chemistry,* the entry-level chemistry course available for students interested in STEM fields or pre-health coursework prior to the curricular revision of Fall, 2009. In contrast, students in Group 2 matriculated in Fall, 2009 or Fall, 2010 and began their chemistry study with Chem20D, *Introduction to Chemistry and Chemical Problem Solving*, the new foundational entry-level chemistry course available for students interested in STEM and health-related fields after the curriculum revision.

In the analysis of grades, only the grade for each student's first attempt in a course was included in the study; subsequent grades earned by any student who repeated after earning a D, F, or W grade on their first attempt were not counted to avoid double counting results for students who withdrew on the first attempt and then failed on the second attempt, for example. Any grades with plus or minus designations were grouped in with the grades that did not contain a plus or minus. As an illustration, B+ and B- grades were placed in the same group with B grades. Grades of D, F, or W were combined into a single category to mimic the wide reporting of DFW rates in the literature. Letter grades were then converted to numerical grades using the following scale: A = 4, B = 3, C = 2, DFW = 1.

Differences in SAT scores and course grades between the two groups were assessed by Student's *t*-tests; a *p* value of < 0.05 was considered statistically significant.

Courses and Programs Prior to the Curriculum and Pedagogy Revisions

Other than the somewhat unusual twist that students are taking chemistry prior to any gateway biology coursework, Duke University's chemistry curriculum was very traditional until the end of the 2007-08 academic year. The four main gateway chemistry courses, two semesters of general chemistry followed by two semesters of organic chemistry, were taught by faculty as large, lecture-style classes. During the fall and spring terms, the total time in lecture was 150 minutes per week spread over either two or three class periods. Each of these courses also contained a weekly three hour lab and 50 minute recitation taught by teaching assistants. Table 1 shows the courses offered in the normal course sequence for all students who entered without advanced placement credit. Aside from the special cases of students with AP credit, this course sequence was the same for all students regardless of whether or not they had taken high school chemistry.

Table 1. Chemistry Course Sequence Experienced by Group 1 (the control group)

Semester (term)	Course number	Course Title
1	Chem21L	General Chemistry
2	Chem22L	General Chemistry
3	Chem151L	Organic Chemistry
4	Chem152L	Organic Chemistry

Several institutional programs and resources were available for students seeking help prior to the curriculum changes. In addition to instructor office hours and department-sponsored resource rooms, the Academic Resource Center (ARC) provided free, one-on-one peer tutors for general and organic chemistry. Tutors would meet weekly with students at any student's request in a massive peer tutoring program. In addition, the ARC ran several peer or staff-led study groups; these were generally offered to students only after they were identified as struggling with coursework through mid-term grades, final grades in a previous term, or advisor or faculty recommendation. Finally, ARC learning specialists offered one-on-one sessions in an academic skills instructional program with a focus on individual student development and self-regulated learning strategies. In addition, the Minority Association of Pre-medical Students (MAPS) organized a learning community each term. This provided peer-led study groups for interested students in organic chemistry.

Courses and Programs after the Curriculum and Pedagogy Revisions

The chemistry curriculum revision involved the addition of a new course for students with relatively less experience in chemistry and chemical problem solving and a realignment of material sequence in the other courses. The new curriculum was implemented for the first time in the 2009-2010 academic year.

Table 2 shows the recommended course sequence for students who matriculate with an SAT math score of 630 or below and/or less than one year of high school chemistry. Since then the courses have been renumbered (Table 2, third column) as part of an institution-wide change, but the content and sequence of courses has not been substantively changed. Students with at least one year of high school chemistry and a higher SAT math score are either advised to start in the second course in the sequence or, if they have a qualifying AP chemistry score of 4 or 5, granted AP credit for one or more courses.

Table 2. Chemistry Course Sequence after the Curriculum and Pedagogy Revisions Experienced by Group 2

Term	Course number	New Number	Course Title
1	Chem20D	Chem99D	Introduction to Chemistry & Chemical Problem Solving
2	Chem31L	Chem101DL	Core Concepts in Chemistry
3	Chem151L	Chem201DL	Organic Chemistry
4	Chem152L	Chem202L	Organic Chemistry
5	Chem32L	Chem210L	Modern Application of Chemical Principles

Introduction to Chemistry and Chemical Problem Solving

The new course, labeled as Chem20D throughout the remainder of this chapter, constituted the department's first effort to replace lecture with active, small group learning sessions in an introductory course. The course was initially formally set up to have one day per week of lecture followed by two days per week of active learning. For the active learning sessions, students worked in small groups on activities drawn from the Process Oriented Guided Inquiry Learning (POGIL) (*36*), Student Centered Activities for Large Enrollment Undergraduate Programs (SCALE UP) (*37, 38*), and problem manipulation (*39*) traditions while the professor and teaching assistants circulated to engage in the discussions. In this way, students were involved in interactive construction of their knowledge while engaging in dialogue with both peers and the professor. When students in a group have more than one idea about how to begin solving a particular problem, the resulting collaborative discourse (argument) increases conceptual understanding by allowing students to effectively resolve wrong claims (*40*) and more closely mirrors the common practice of argument and debate of professional scientists (*41*). The one day per week formerly consisting of lecture has now been be web-enhanced: students can watch several 10-20 minute online lecture videos per week. These videos have integrated interactive questions embedded every few minutes to improve attention and comprehension (*42*). Class time formerly devoted to live lectures has been replaced by an instructor-led in-class discussion and problem solving session. The course has pre-determined percentage cut-offs for letter grades published in the syllabus with the goal of encouraging students to collaborate in their learning rather than compete as they sometimes do in situations where grades are assigned by curving.

In order to prevent students from enrolling in a more elementary course than their background warrants, Chem20D does not contain a lab and does not count toward a chemistry major, chemistry minor, or pre-medical requirements. It is truly intended to be a beginning course in chemistry, but, similar to a college beginning foreign language class, it is not a remedial class. Chem31L instructors, in contrast, can now more realistically assume a certain baseline knowledge of fundamental chemistry for students in that class. So, basic concepts such as stoichiometry and compound naming are not covered in Chem31L. Instead, this class begins with thermochemistry, assuring that students will not be bored or lulled into a false sense of security and develop the poor study and class attendance habits that sometimes accompany such boredom. Aside from Chem20D, the other courses remained in lecture/lab/recitation format initially, although about one third of the sections of Chem31L and Chem151L have now converted to more student-centered formats drawing from Team-Based Learning and POGIL literature, respectively. At the opposite edge of the spectrum from Chem20D, students with more extensive pre-college chemistry coursework and an AP score of 4 or 5 would be unchallenged in Chem20D or Chem31L, so they place into Chem110L (*Honors Chemistry*: a single semester of accelerated general chemistry) or Chem151L.

At the same time that the curriculum and pedagogy changes occurred in chemistry, both the pre-major academic advising group and the ARC implemented new learning communities for students: the Cardea Fellows (*43*) and the Science Advancement through Group Engagement (SAGE) programs, respectively (*44*). These programs initially focus on inviting the students who enroll in Chem20D to participate, and results of pilot studies of these programs have been previously described (*43, 44*).

Limitations

Before beginning the discussion of results, a few limitations of this work should be noted. This study was relatively small in scope and sample size and was conducted in an American educational context. The undergraduate student population at Duke University is fairly homogeneous in terms of age; >90% of enrolled undergraduate students are traditional college age (under 24 years old). In addition, because Duke University is a highly selective institution in terms of undergraduate admissions, the students in this study may not constitute a representative sample of the global population of all students planning to pursue STEM or health professions. The students in the study were full-time students, so no examination of enrollment intensity (whether students are studying part-time or full-time) can be conducted. Most students in this study neither worked full-time during the academic year nor served as the primary caregiver of children. Due to the characteristics of the populations studied herein, caution must be exercised in attempting to extrapolate the findings to populations in which a large number of students work full time and/or serve as the primary caregiver to children. Because the revised curriculum offers Chem20D in the fall semester only, the experimental group was slightly smaller than the control group, which includes students who took their first semester of Chem21L in either the fall or spring terms prior to the curriculum revision. Because students took the courses in different terms and in many cases with different instructors, grade distributions are not identical for each course, but each sample analyzed contains grades from two or more semesters, and the gross differences in outcomes observed in this study cannot likely be explained by small deviations in grade distributions from semester to semester. Students who matriculated into the Pratt School of Engineering were not included in this study, nor were students ($N = 9$) who did not have an SAT math score (or ACT math score equivalent) reported in the records database. To protect individual confidentiality and due to the small number of students in some groups, some data could not be disaggregated further by race and ethnicity.

Results and Discussion

First, the SAT math scores of the two groups were compared: Group 1 had a 20 point higher average SAT math score than Group 2, and this result was statistically significant ($p < 0.001$). This most likely arose because Group 2 did *not* contain the students who chose to enroll in Chem31L as their first college chemistry course. This latter group of students, who self-selected out of Group 2, tended to have more experience in both pre-college math and science and slightly higher SAT math scores than the students who enrolled in Chem20D.

Next, the effectiveness of the previous curriculum and exposition-style coursework was explored for the students in Group 1 through an examination of grade outcomes in their first general chemistry course (Chem21L). Similarly, the effectiveness of the revised curriculum, which included a new introductory course and more student-centered coursework, was explored for the students in Group 2 by an examination of their grade outcomes in the new introductory course (Chem20D) and the revised general chemistry course (Chem31L). Figure 1 contrasts the first term grades earned by the students in these groups in chemistry classes before and after the curriculum revision. The distribution of first semester chemistry grades for students in the lowest quartile of math SAT scores changed dramatically, the observed differences were highly statistically significant ($p < 0.001$), and this change in student success correlated with the institutional changes.

For reference, both before and after the curriculum revision, a typical overall general chemistry class grade distribution at our institution contains more than 25% A grades. As an illustration, the two combined sections of Chem21L in Fall, 2007 ($N = 434$), which was before the curriculum revision, had the following letter grade distribution: 26.5% A, 35.5% B, 28.6 % C, and 9.4% DFW. While these overall class grade distributions might look fairly generous to readers who teach college general chemistry, Figure 1 shows that the situation was actually quite bleak for the students in that course who were in the lowest quartile of SAT math scores (Group 1): over 75% of those who enrolled in Chem21L in the years which were data mined received a grade of C, D, F, or W. Moreover, students in Group 1 earned *an order of magnitude fewer* A grades than their classmates.

Several learning resources were available for Group 1, as described previously, and these generally required some level of independent student help-seeking. Low grades earned by this group indicate that the learning resource system operating at that time was not effective in promoting the achievement of high tests scores for the majority the students in Group 1. We do not know if the resources were inadequate, if students did not seek them out, or both. There is evidence for both in the literature: Arroyo notes that "help seeking by itself is not sufficient to achieve learning (*45*)," and help seeking avoidance patterns are commonly observed in students in large introductory chemistry classes (*46*).

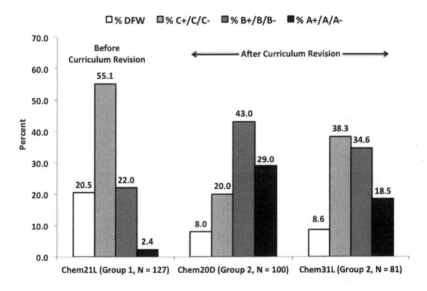

Figure 1. First year chemistry course grade outcomes for students in the lowest 25th percentile of SAT math scores for their matriculating class.

The curriculum changes, implementation of Chem20D, and development of new learning communities correlated to an improvement in outcomes for the students in Group 2 compared to those in Group 1 as shown clearly in Figure 1. The Chem20D course focused on developing a firm foundation of knowledge in chemical concepts and competence in chemical problem solving within an active learning classroom environment. Active and student-centered instruction have since been shown to be more efficacious than lecturing for STEM learning in both a double blind study (*47*) and a large metaanalysis (*48*). In addition, the Chem20D group-learning structure and development of new learning communities provided a framework for the formation of a culture of scholarship and engagement in team learning both inside and outside of the formal classroom setting (*49*). Student academic engagement is known to be essential to deep learning in college STEM coursework (*50*), so promotion of practices that lead to higher levels of engagement was a focus of these institution-wide efforts.

Since practitioners debate the value of remediation (*51*), it is important to note that this course is not considered to be remedial any more than a beginning course in Spanish or another foreign language might be considered to be remedial. Most incoming high school students have one year or less of chemistry courses, and these students are still beginners in the field. This contrasts sharply with remedial college math classes because most undergraduates matriculate with at least of decade of math courses.

Some researchers have hypothesized and found evidence that any benefits of preparatory courses quickly "fade out," with the initially observable effect not lasting until the second or third semester after treatment (*52*). To test this hypothesis, student outcomes in organic chemistry were assessed. For both groups of students, the organic chemistry 1 class is the third chemistry class in their normal curricular sequence. Students in Group 2 had more A grades and fewer DFW grades than students in Group 1 (Figure 2), but the overall difference in grades was not statistically significant (*p* = 0.059).

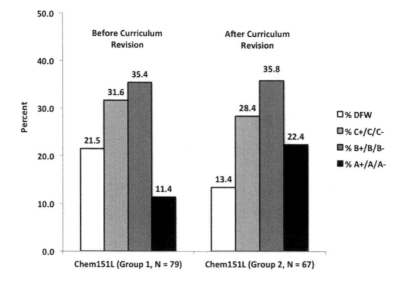

Figure 2. Organic chemistry course grade outcomes for students in the lowest 25th percentile of SAT math scores for their matriculating class before and after the curriculum revision.

The effect of pedagogy changes in isolated cases of one or two gateway courses has shown mixed results in the literature. Watkins and Mazur implemented peer instruction with ConcepTests in an introductory physics course and found "compelling evidence that a single course can have a significant long-term impact on the retention of students in STEM majors (*53*)." In contrast, Lewis found changing only general chemistry courses to be inconsequential in terms of retention in the chemistry major, leading to a call for "implementing and evaluating curricular-wide reform (*54*)." It is difficult to say at this early point in the analysis whether or not our results are more in line with the first report (*53*), but it must be noted that the two institutions (Harvard in the study by Watkins and Mazur versus Duke in the study reported here) are similar. For example, both institutions have highly selective undergraduate admissions and are comprised predominantly of students who are traditional college age and do not face the pressures of full time employment or raising children. It is possible that students in the environment studied by Lewis (*54*). need more active style coursework throughout the curriculum to stay successful, while students who face fewer non-academic challenges might only need positive experiences in the first science course they take. Further study is needed in this area.

Demographic Implications

Although numerous reports point to relatively low diversity in groups of graduating college seniors who earn degrees in STEM fields (*2, 8*) and a corresponding dearth of URM professionals in engineering and the physical sciences (*55*), large numbers of URM students enter college with the intention of studying in science, engineering, and health tracks. With this in mind, aggregated self-reported demographics for the groups studied herein are described in Table 3. The numbers do not add up to 100% because a small number of students did not indicate race, and the Native Americans were not broken out as a separate group due to the small numbers of students.

Table 3. Description of Sample

Race/Ethnicity	Group 1	Group 2
Asian/Asian American	6%	5%
Black/African American	38%	43%
Hispanic/Latino	17%	19%
White/Causasian	36%	29%
Gender		
Male	26%	25%
Female	74%	75%

The gender balance was skewed towards females in both groups; Group 2 contained ~7% more URM students than Group 1. More importantly, both groups contained a disproportionately high representation of women and URM students when compared to the overall composition (*56*) of the undergraduate student body, which is 50% female and 18% URM and has been this approximate composition for more than a decade.

To understand why there is over-representation of women and URM students in these groups, one needs to consider that national data reveal disparities in SAT scores as a function of race, ethnicity, gender, and family socioeconomic status (*57, 58*). For example, the composite SAT scores for students who said their families earned over $200,000 per year was nearly 400 points higher than scores for students with family incomes of $20,000 per year, and male students still consistently score higher on the SAT than female students. So, the overrepresentations of URM and female students in the groups studied here who have relatively low SAT scores simply mirror trends seen on a national scale.

Conclusions and Future Directions

The disparities between student interests and outcomes lead to important questions: do the higher rates of attrition of URM students arise from insurmountable deficits in learner backgrounds or psychosocial factors such as stereotype threat? Or does the finding arise primarily from a lack of both institutional readiness and appropriate programming for teaching a broadly diverse student body? Results in this study indicate that the latter likely accounts for much of the observed differences in STEM graduation rates at highly selective institutions. Three strategies were employed to improve outcomes for students near the lower edge in terms of prior science experience: curriculum revisions, implementation of more evidence-based teaching practices, and creation of communities of STEM scholarship both inside and outside of the classroom. STEM major and minor graduation rates for URM students are just now beginning to increase as a result of institutional changes outlined here. The Trinity College Office of Assessment and the Department of Chemistry at Duke University continue to work in partnership on a longitudinal study that uses a multivariable regression model to study this phenomenon for a much larger sample over a longer period of time.

Acknowledgments

Dick MacPhail and the late Jim Bonk are thanked for their leading roles in developing the new chemistry curriculum and placement guidelines. Dick MacPhail is thanked for many fruitful pedagogical discussions, generous sharing of data, and manuscript review. Miguel Bordo, Nyote Calixte, Andrea Novicki, Lee Willard, and the book's editors are also thanked for manuscript review. Keith Whitfield and the members of the "flipped classroom group" are thanked for on-going support in the study of classroom pedagogy efficacy. Andrea Novicki and Randy Riddle are thanked for their efforts in leading the 2013-14 *Flipped Faculty Scholars* group. Lee Baker and Alyssa Perz-Edwards are thanked for setting up the Cardea Fellowship Program and supporting the SAGE program. ARC personnel are thanked for their collaboration in the creation and maintenance of the SAGE program. Jennifer Hill is thanked for more recent data mining efforts and collaboration in the ongoing longitudinal analysis.

A portion of the work described herein was supported by an on-going grant from the Howard Hughes Medical Institute. Resources to assess the curriculum and pedagogy changes came from a grant by the Duke University Arts and Sciences Faculty Assessment Committee. Financial support for piloting student-centered activities was provided by a Paletz Innovative Teaching grant,

Dean of Arts and Sciences' mini-grant, Center for Instructional Technology's faculty fellowship, and the Provost's Office Flipped Classroom Committee.

References

1. Malcom-Piqueux, L. E.; Malcom, S. M. Engineering diversity: fixing the educational system to promote equity. *The Bridge* **2013**, *43*, 24–34.
2. Chen, X. *STEM Attrition: College Students' Paths Into and Out of STEM Fields*; NCES 2014-001; National Center for Education Statistics, Institute of Education Sciences, U.S. Department of Education: Washington, DC, 2013.
3. Daempfle, P. A. An analysis of the high attrition rates among first year college science, math, and engineering majors. *J. Coll. Stud. Ret.* **2004**, *5*, 37–52.
4. Allen-Ramdial, S. A.; Campbell, A. G. Reimagining the pipeline: advancing STEM diversity, persistence, and success. *BioScience* **2014**, *64*, 612–618.
5. President's Council of Advisors on Science and Technology, *Engage to Excel: Producing One Million Additional College Graduates with Degrees in Science, Technology, Engineering, and Mathematics*; The White House: Washington, DC 2012.
6. Hrabowski, F. A., III. Broadening participation in the American STEM workforce. *BioScience* **2012**, *62*, 325–326.
7. Hrabowski, F. A., III. Boosting Minorities in Science. *Science* **2011**, *331*, 125.
8. Garrison, H. Underrepresentation by race-ethnicity across stages of U.S. science and engineering education. *CBE Life Sci. Educ.* **2013**, *12*, 357–363.
9. Heavy, S. More Minority Babies Than Whites in U.S.: Census Bureau. *Chicago Tribune* **2012**, May 17, 2012.
10. Barr, D. A.; Matsui, J.; Wanat, S. F.; Gonzalez, M. E. Chemistry as the turning point for premedical students. *Adv. Health Sci. Educ. Theory Pract.* **2010**, *15*, 45–54.
11. Barr, D. A. *Questioning the Premedical Paradigm: Enhancing Diversity in the Medical Profession a Century after the Flexner Report*; Johns Hopkins University Press: Baltimore, MD, 2010; pp 11–34.
12. Chang, M. J.; Sharkness, J.; Hurtado, S.; Newman, C. B. What matters in college for retaining aspiring scientists and engineers from underrepresented racial groups. *J. Res. Sci. Teach.* **2014**, *51*, 555–580.
13. Labov, J. B.; Singer, S. R.; George, M. D.; Schweingruber, H. A.; Hilton, M. L. Effective practices in undergraduate STEM education part 1: examining the evidence. *CBE Life Sci. Educ.* **2009**, *8*, 157–161.
14. Winkleby, M. A.; Ned, J.; Ahn, D.; Koehler, A.; Kennedy, J. D. Increasing diversity in science and health professions: a 21-year longitudinal study documenting college and career success. *J. Sci. Educ. Technol.* **2009**, *18*, 535–545.
15. Kuh, G. D. *High-Impact Educational Practices: What They Are, Who Has Access to Them, and Why They Matter*; AAC&U: Washington, DC, 2008.

16. Hughes, A. N.; Gibbons, M. M.; Mynatt, B. Using narrative career counseling with the underprepared college student. *Career Development Quarterly* **2013**, *61*, 40–49.

17. Shields, S. P.; Hogrebe, M. C.; Spees, W. M.; Handlin, L. B.; Noelken, G. P.; Riley, J. M.; Frey, R. F. A transition program for underprepared students in general chemistry: diagnosis, implementation, and evaluation. *J. Chem. Educ.* **2012**, *89*, 995–1000.

18. Hock, M. F.; Deshler, D. D.; Schumaker, J. B. Tutoring programs for academically underprepared college students: a review of the literature. *J. Coll. Read. Learn.* **1999**, *29*, 101–122.

19. Reyna, C. Ian is intelligent but Leshaun is lazy: antecedents and consequences of attributional stereotypes in the classroom. *Eur. J. Psychol. Educ.* **2008**, *23*, 439–458.

20. Miller, B. K. Measurement of academic entitlement. *Psychol. Rep. Sociocultural Issues Psychol.* **2013**, *113*, 654–674.

21. Koebler, J. Killing STEM Achievement. U.S. News and World Report, 2012. http://www.usnews.com/news/blogs/stem-education/2012/04/19/experts-weed-out-classes-are-killing-stem-achievement (accessed April 23, 2015).

22. Leslie, S.-J.; Cimpian, A.; Meyer, M.; Freeland, E. Expectations of brilliance underlie gender distributions across academic disciplines. *Science* **2015**, *347*, 262–265.

23. Beasley, M. A.; Fischer, M. J. Why they leave: the impact of stereotype threat on the attrition of women and minorities from science, math, and engineering majors. *Soc. Psychol. Educ.* **2012**, *15*, 427–448.

24. Davis, J. F. The campus as a frog pond: an application of the theory of relative deprivation to career decisions of college men. *Am. J. Soc.* **1966**, *72*, 17–31.

25. Smyth, F. L.; McArdle, J. J. Ethnic and gender differences in science graduation at selective colleges with implications for admission policy and college choice. *Res. High. Educ.* **2004**, *45*, 353–381.

26. Espinosa, L. L. Pipelines and pathways: women of color in undergraduate STEM majors and the college experiences that contribute to persistence. *Harvard Educ. Rev.* **2011**, *81*, 209–240.

27. Chang, M. J.; Cerna, O.; Han, J.; Sàenz, V. The contradictory roles of institutional status in retaining underrepresented minorities in biomedical and behavioral science majors. *Rev. High. Educ.* **2008**, *31*, 433–464.

28. Smit, R. Towards a clearer understanding of student disadvantage in higher education: problematising deficit thinking. *High. Educ. Res. Dev.* **2012**, *31*, 369–380.

29. Rose, T. *The Myth of Average*, 2013, TedxSonomaCounty. http://tedxtalks.ted.com/video/The-Myth-of-Average-Todd-Rose-a (accessed April 23, 2015).

30. Weber, R. N. Manufacturing gender in commercial and military cockpit design. *Sci. Tech. Human Values* **1997**, *22*, 235–253.

31. Stiehm, J. Women's Biology and the U.S. Military. In *Women, Biology, and Public Policy*; Sapiro, V., Ed.; Sage: Beverly Hills, CA, 1985; pp 205–234.

32. Jeffrey, L. M. Learning orientations: diversity in higher education. *Learn. Ind. Differences* **2009**, *19*, 195–208.

33. Grove, N. P.; Bretz, S. L. A continuum of learning: from rote memorization to meaningful learning in organic chemistry. *Chem. Educ. Res. Pract.* **2012**, *13*, 201–208.

34. Quinnell, R.; May, E.; Peat, M. Conceptions of biology and approaches to learning of first year biology students: introducing a technique for tracking changes in learner profiles over time. *Int. J. Sci. Educ.* **2012**, *34*, 1053–1074.

35. Grove, A. *Converting ACT Scores to SAT Scores*. http://collegeapps.about.com/od/standardizedtests/a/convertSAT2ACT.htm (accessed April 23, 2015).

36. *POGIL: Process Oriented Guided Inquiry Learning*; Moog, R. S., Spencer, J. N. Eds.; Oxford University Press: New York, 2008.

37. M. T. Oliver-Hoyo, R. J. Beichner, SCALE-UP: Bringing Inquiry-Guided Learning to Large Enrollment Courses. In *Teaching and Learning through Inquiry*; Lee, V. S., Ed.; Stylus: Sterling, VA, 2004; pp 51−79.

38. Oliver-Hoyo, M. T. Content Coverage in a Lecture Format versus Activity-Based Instruction. In *Investigating Classroom Myths through Research on Teaching and Learning*; Bunce, D. M., Ed.; ACS Symposium Series 1074; Americal Chemical Society: Washington, DC, 2011; pp. 33−50.

39. Siburt, C. J. P.; Bissell, A. N.; MacPhail, R. A. Developing metacognitive and problem-solving skills through problem manipulation. *J. Chem. Educ.* **2011**, *88*, 1489–1495.

40. Kulatunga, U.; Moog, R. S.; Lewis, J. E. Argumentation and participation patterns in general chemistry peer-led sessions. *J. Res. Sci. Teach.* **2013**, *50*, 1207–1231.

41. Osborne, J. Arguing to learn in science: the role of collaborative, critical discourse. *Science* **2010**, *328*, 463–466.

42. Szpunar, K. K.; Khan, N. Y.; Schacter, D. L. Interpolated memory tests reduce mind wandering and improve learning of online lectures. *Proc. Nat. Acad. Sci.* **2013**, *110*, 6313–6317.

43. Perz-Edwards, A. *The Cardea Fellows Program: Supporting Science Learning in the Prehealth Curriculum*. Paper presented at the 20th National NAAHP Meeting, Baltimore, MD, June 23, 2012.

44. Hall, D. M.; Curtin-Soydan, A. J.; Canelas, D. A. The science advancement through group engagement program: leveling the playing field and increasing retention in science. *J. Chem. Educ.* **2014**, *91*, 37–47.

45. Arroyo, I. *Inferring Unobservable Learning Variables from Students' Help Seeking Behavior*; Computer Science Faculty Publication Series; 2004; paper 110;http://scholarworks.umass.edu/cgi/viewcontent.cgi?article=1109&context=cs_faculty_pubs (accessed April 23, 2015).

46. Karabenick, S. A. Perceived achievement goal structure and college student help seeking. *J. Educ. Psychol.* **2004**, *96*, 569–581.

47. Granger, E. M.; Bevis, T. H.; Saka, Y.; Southerland, S. A.; Sampson, V.; Tate, R. L. The efficacy of student-centered instruction in supporting science learning. *Science* **2012**, *338*, 105–108.

48. Freeman, S.; Eddy, S. L.; McDonough, M.; Smith, M. K.; Okoroafor, N.; Jordt, H.; Wenderoth, M. P. Active learning increases student performance in science, engineering, and mathematics. *Proc. Nat. Acad. Sci.* **2014**, *111*, 8410–8415.

49. Varma-Nelson, P.; Coppola, B. P. Team learning. In *Chemist's Guide to Effective Teaching*; Pienta, N; Cooper, M. M.; Greenbowe, T., Eds.; Pearson: Saddle River, NJ, 2005; pp. 155–169.

50. Gasiewski, J. A.; Eagen, M. K.; Garcia, G. A.; Hurtado, S.; Chang, M. From gatekeeping to engagement: a multicontextual, mixed method study of student academic engagement in introductory STEM courses. *Res. High. Educ.* **2012**, *53*, 229–261.

51. Long, B. T. The remediation debate. *National CrossTalk* **2005**, *13*, 11–12.

52. Garland, E. R.; Garland, H. T. Preparation for high school chemistry: the effects of a summer school course on student achievement. *J. Chem. Educ.* **2006**, *83*, 1698–1702.

53. Watkins, J.; Mazur, E. Retaining students in science, technology, engineering, and mathematics (STEM) majors. *J. Coll. Sci. Teach.* **2013**, *42*, 36–41.

54. Lewis, S. E. Investigating the longitudinal impact of successful reform in general chemistry on student enrollment and academic performance. *J. Chem. Educ.* **2014**, *91*, 2037–2044.

55. Bayer Facts of Science Education XIV: Female and Minority Chemists and Chemical Engineers Speak about Diversity and Underrepresentation in STEM, March, 2010. http://bayerfactsofscience.online-pressroom.com/#a (accessed April 23, 2015).

56. *Quick Facts About Duke*. Office of News and Communication, Duke University. http://newsoffice.duke.edu/all-about-duke/quick-facts-about-duke (accessed April 23, 2015).

57. Marklein, M. B. SAT Scores Show Disparities by Race, Gender, Family income. *USA Today*; August, 26, 2009; Gannett Company: Fairfax, VA, 2009.

58. Jaschik, S. SAT Scores Drop Again. Inside Higher Ed, September 25, 2012. https://www.insidehighered.com/news/2012/09/25/sat-scores-are-down-and-racial-gaps-remain (accessed April 23, 2015).

Chapter 3

The Flipped Classroom as an Approach for Improving Student Learning and Enhancing Instructor Experiences in Organic Chemistry

Thomas Poon*,[1] and Jason Rivera*,[2]

[1]Professor of Chemistry, W.M. Keck Science Department,
Claremont McKenna College, Pitzer College, Scripps College,
925 N. Mills Ave., Claremont, California 91711, United States
[2]Director of Institutional Research, Office of Institutional Research,
Dickinson College, P.O. Box 1773,
Carlisle, Pennsylvania 17013, United States
*E-mail: tpoon@kecksci.claremont.edu (T.P.);
jason.e.rivera@outlook.com (J.R.)

The flipped classroom is a form of hybrid instruction in which active forms of engagement inside the classroom are made possible by the delivery of course content outside of the classroom, usually in the form of videos or other digital media. While the benefits of active learning in the classroom have been widely reported in the literature, less emphasis has been placed on the development of video instruction and the pedagogical advantages that this medium provides. This chapter describes strategies for enhancing video instruction and for coupling it with active classroom-based pedagogies in a flipped classroom approach to the yearlong organic chemistry course. These strategies extend student accessibility to course content, improve student learning, and provide instructors with opportunities to enhance their teaching and research portfolios.

Introduction

In the traditional brick and mortar approach to teaching, classroom lectures often represent the students' first exposure to the course content, or at the very least, an opportunity for the instructor to impart his or her wisdom and perspective on the topic at hand. Here, time spent outside of the classroom usually emphasizes student processing of the material through activities such as problem sets, writing assignments, and assigned readings. In the flipped classroom approach (1–6), the student's first exposure to the course content is achieved through video lectures, while class time is devoted to an active form of pedagogy such as working in groups, class discussions, or the use of clickers. The flipped classroom has been lauded (7, 8)for its ability to shift the processing of course material to an active mode led by the instructor, thereby enhancing the in-class student experience and student learning overall.

The benefits of active pedagogies have been amply reported in the literature (9–12), and this is perhaps the reason the classroom component of the flipped approach has garnered the most attention in educational circles (13–15). The video instructional component, on the other hand, has not received as much attention, and flipped classroom adopters are often advised to focus more on the in-class student learning experience (16). This need not be the case, and in fact, would represent a lost opportunity to enhance student learning.

Tangible Advantages of Video Instruction

Video instruction has been implemented in science education since the technology first became available (17), and its uses have been varied. For example, videos have been used to prepare and support students for specific tasks such as using chemical instrumentation or setting up experiments (18, 19). Videos can allow students to make up missed content from absences, such as when they join a course after the term has begun (e.g., during the "shopping period" that students at many institutions practice) or when there are prolonged absences due to illness or other exigent circumstances. Videos have also been used by instructors to assess their own teaching. Most recently, with the advent of MOOCs (20–22) and other sources of online instruction such as Khan Academy (23), videos have been used to teach new material and entire courses.

Videos represent a student-centered form of pedagogy because they allow students to learn at their own pace, to access the content when it is most convenient to them, and to review the content as often as needed. Videos can also be structured to address the limited attention span inherent in most learners (24) by (1) splitting video lectures into separate, shorter video files or (2) by designing videos with timesaving strategies in mind. The latter could be achieved by creating the video such that drawings and text appear instantaneously on the screen (Figure 1) or by editing out superfluous portions of the video.

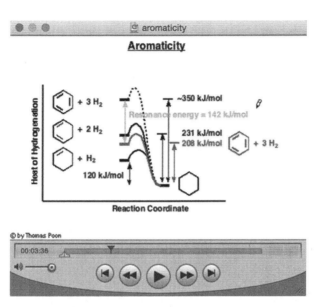

Figure 1. Representative video screen shot showing elements that are predrawn such that their appearance on screen is instantaneous, allowing for timesaving and shortening of video duration.

Critics of asynchronous online instruction have often cited its passive nature as a shortcoming of the medium (*25, 26*). In the experience of one of the authors (Poon; herein referred to as "this author"), active learning can be injected into video instruction simply by prompting students to pause the video, attempt the on-screen problem on their own, and subsequently restart the video to see how the problem is solved (Figure 2). This pause-continue strategy is similar to the Predict-Observe-Explain strategy (*27*) used by Kearney *et. al.* and could be used in various other ways to make learning active for students. For example, students could be entreated to use the internet to gather data or information, to complete a reading assignment before proceeding, or to work on a task with other students.

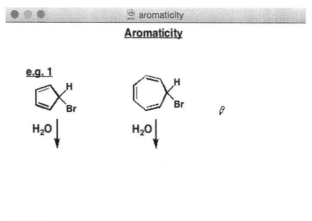

Figure 2. Representative video screen shot showing a moment of active learning. Students are instructed to pause the video, attempt the problem on their own (in this example, comparing the rates of reaction in the E1 reactions shown), and restart the video to see how the problem is solved.

Another criticism of online instruction is that students, unless they are interacting in a synchronous video feed, are unable to have their questions immediately addressed by the instructor (*28*).   While this limitation of asynchronous video instruction is valid, technology can certainly facilitate the fielding of student inquiries.   In this author's course, students view the video lectures in the foreground, while a browser containing a form for submitting questions resides in the background (Figure 3).   In this manner, each student is able to pose questions to the instructor, and these queries could be answered by the instructor or by teaching assistants.   This approach typically garners more queries than would be received in the classroom and promotes participation from students who might not otherwise ask a question in class.   It also has the benefit of allowing the instructor to tailor his or her classroom activity in response to the questions and issues raised by students. This ability to be dynamically responsive to the general class understanding (or misunderstanding) of the subject matter is a benefit that is more difficult to achieve in most lecture-based classroom environments.

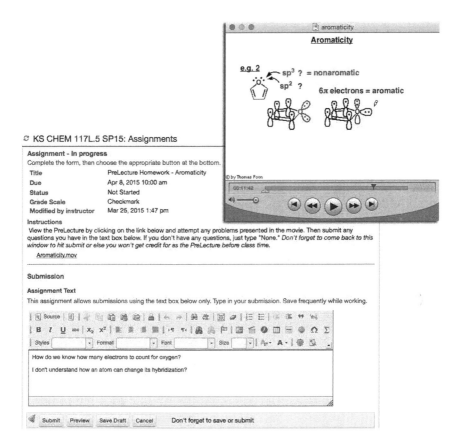

Figure 3. Representative asynchronous interface for facilitating student questions during review of online course videos. The video lecture screen shot is viewed in the foreground, while a text input window is available in the background. (Used with permission from the Sakai course management system.)

The Flipped Classroom in Organic Chemistry – One Instructor's Approach

Since September 2006, this author has utilized the flipped classroom approach in his two-semester organic chemistry course. Currently, the course is such that 68% of class meetings in the first semester are flipped and 49% of the classes in the second semester are flipped. Here, "flipped" is used to specify class meetings in which student viewing of the online lectures is required beforehand in order to take full educational advantage of the in-class activity. From day one of the course, the following policies and practices are established:

- Students are assigned homework, which consist solely of viewing video lectures called *PreLectures*. The assignment due dates are listed for the entire semester both on the syllabus and on the course management software (CMS) website.

- PreLectures are made available on the CMS as the semester progresses, rather than all at once, usually 2-4 days prior to the class in which they are due. Student viewing of the PreLectures is factored into the class participation grade, which is 2% of the overall grade in the course. Tracking of PreLecture views is done via the CMS.
- Many PreLectures are designed to be interactive (Figure 2) and the CMS interface allows for student questions to be sent to the instructor via email (Figure 3). The instructor responds to all emailed questions within 24 h.
- Time spent in the classroom is mostly focused on the use of personal response systems (a.k.a. clickers), but also includes discussions of organic chemistry in the news, problem solving in small groups, going over recent exams, and a presentation on summer research opportunities.

Comparison of the Flipped versus Non-Flipped Classroom

In the fall of 2009, this author taught consecutive sections of organic chemistry, and was able to teach one section using the flipped approach and a second section such that the students' first exposure to course content was via traditional lectures. Clickers were used in both classes, but the non-flipped section did not have the benefit of group problem solving activities nor did it receive the in-class presentation on summer research opportunities (the opportunities were provided as a handout instead). The non-flipped section also saw much less coverage of organic chemistry in the news. One other difference between the two sections was the number of clicker questions covered. The flipped classroom was presented with 109 clicker questions, while the non-flipped classroom attempted 78 clicker questions (28% fewer clicker questions).

There were, of course, similarities between the two sections. Both courses utilized the same textbook, shared the same office hours and review sessions, had the same schedule of topics on their syllabi, and evaluated students using identical exams. The examinations were given within 10 minutes of each other and in rooms on campus separated by over 200 yards, which presumably prevented students in the earlier section from influencing students in the later section. The videos were also made available to the students in the non-flipped section through the website http://www.ochem.com. However, at this public website, the videos are not linked to the course syllabus in any way and students in the non-flipped class were neither required nor asked to view the videos prior to the corresponding class session.

The flipped approach to organic chemistry described in the previous section and the comparison study done in 2009 revealed much about the advantages of the flipped classroom pedagogy. It should be noted, however, that at this author's institution, small class sizes are the norm and organic chemistry classes range from 24–40 students each semester. In the comparison study done in fall 2009, there were 31 students enrolled in the flipped class and 33 students in the non-flipped class. A series of t-tests were performed for all exams to determined whether the flipped section performed significantly better than the non-flipped section, especially with relation to the final exam.

Table 1 shows data for the two classes taught by this instructor in 2009. The first and second rows show the mean scores for the non-flipped class and the flipped class, respectively. The third and fourth rows show data for the flipped class only. Results indicated a non-significant difference trending in the anticipated direction between the flipped section (M=77.1; SD=13.9) and non-flipped section (M=72.6; SD=15.2) for final exam performance, $t(51)$= -1.14, p=0.13. To understand why the final exam scores may be trending towards being significantly higher for the flipped class, the percentage of flipped classes leading up to each exam and the percentage of students viewing the PreLectures were calculated. It would be interesting to see, in the case of any future studies like this, whether having more flipped classes and increasing student pre-viewing of the videos would significantly affect differences in student learning. It is important to note that the flipped class did not perform worse than the students in the traditionally taught class.

Table 1. Average Exam Performance and Flipped Classroom Data for Two Fall 2009 Organic Chemistry Sections

	Mini-exam*	Exam 1	Exam 2	Exam 3	Final
Non-flipped section (%)	77	76	79	76	73
Flipped section (%)	79	78	82	77	77
Percentage of flipped classes since the previous exam that led up to each exam	86	100	100	56	82
Percentage of students in the flipped course viewing videos prior to attending class for that exam period	70	70	47	47	58

* A 25 min. exam that covered the first chapter of the course.

Another point of comparison are the instructor teaching evaluations conducted for the two 2009 courses. Students rated categories for the flipped class on par with or better than that for the non-flipped class. For example, when surveyed about the instructor's "use of class time," students rated the flipped class at 5.71 and the non-flipped class at 5.48 (six-level Likert scale). For the question, "How does this course compare to other courses offered you have taken at The Claremont Colleges?" the ratings were 5.74 (flipped) and 5.15 (non-flipped). One additional example is the rating of the "instructor's effectiveness in teaching the subject matter;" 5.87 (flipped) versus 5.73 (non-flipped).

Longitudinal performance data for this author's spring semester final examinations since 2005 are shown in Figure 4. A two-sample unequal variance t-test was performed on the average ACS score over the two years when the course was taught using the traditional approach and the average ACS score over the five years when the course was taught using the flipped approach. The t-test demonstrates a statistically significant difference between flipped approach scores (M=56.5; SD=9.3) and non-flipped scores (M=51.3; SD=10.6), $t(193)$= -3.35, p<.001. The longitudinal data clearly supports the benefits of the flipped approach for teaching organic chemistry.

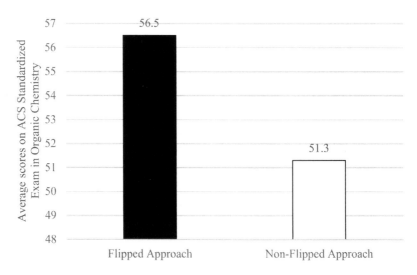

Figure 4. Average student raw scores on the 2004 version of the ACS Standardized Exam in organic chemistry. The average flipped approach score is from exams taken in 2007, 2008, and 2011-2013 (N=150). The non-flipped approach average is from exams taken in 2005 and 2006 (N=57). This author did not teach in the spring semesters of 2009, 2010, and 2014.

Discussion

Meyer, Rose, and Gordon, the authors of *Universal Design for Learning: Theory and Practice (29)*, used research on cognitive neuroscience and learning to suggest that effective curricula must be varied in its approach to delivering course content, skills development, and evaluation in order to reach students and their various learning styles. The authors identified three networks in the brain that are vital for effective learning, isolated their functions, and proposed best practices for effectively accessing these cerebral learning networks in all learners. These items are summarized in Table 2.

In this author's experience with the two-semester organic chemistry course, the flipped classroom approach makes it possible to achieve more of these best practices, which in turn enhance the learning functions of students according to Meyer, *et. al.* The student-centered approach of asynchronous video instruction and the use of clicker questions in the classroom present two very different ways of representing the material to students in order to help them understand and identify the concepts (recognition networks). Using clickers and facilitating the submission of questions and the pause-continue strategy described for the PreLecture videos provide different ways for students to respond and express their knowledge, which in turn help students to plan and execute their learning of the material (strategic networks). The interjection of topical news items and devoting class time to discussing summer research opportunities available to students helps students to engage with their learning (affective networks).

Table 2. Rose and Meyer's Universal Design for Learning (*29*)

Cerebral Network	Function	Best Practices for Engaging
Recognition	Identifying and understanding	Represent the material in multiple ways.
Strategic	Planning and executing	Provide multiple ways for students to respond and to express knowledge.
Affective	Engaging with learning	Provide multiple avenues for interaction, collaboration, and reflection.

Using clickers, conducting small group activities, and spending time on enrichment topics or mentoring activities all use valuable class time that is made possible because the introduction of basic concepts and course material is shifted to the PreLecture videos. In this author's experience, this had the added benefit of always being on schedule with regard to the syllabus. Further, the online presentation of the basic concepts allowed for more in-depth coverage of the course material, such as being able to do 28% more clicker questions in the flipped class than in the non-flipped class (see previous section). The clicker approach allows one to be dynamically responsive to the students' understanding (or lack thereof) of the material. This creates a freshness and unknown aspect to each class that prevents the course from becoming stale. Because the clicker questions have already been developed and the PreLectures do the job of "teaching" the basic concepts, most of an instructor's class preparation time can be devoted to creating a more student-centered course via the answering of student questions by email and the adaptation of material to address student misconceptions.

Strategies for Implementing the Flipped Approach

This author's implementation of the flipped approach was made possible by a grant from the Camille & Henry Dreyfus Foundation to create the video PreLectures as part of their Special Grant Program in the Chemical Sciences, and from an internal institutional grant to purchase the clickers that were used to infuse an active learning pedagogy into the classroom. At the time of the Dreyfus grant in 2004, production of the video tutorials required a powerful computer and several software-based applications. Today, not only are the tools for creating videos more readily available and powerful, but the number of individuals, including faculty, on a college campus who have the skills to develop instructional videos has vastly increased. Because faculty can often readily produce or obtain assistance with producing videos in order to flip their classes, this section will be devoted more to offering strategies, rather than to providing technical advice on the video creation software.

Instructors who wish to flip their classes should first decide on the general format of the videos they wish to produce. For example, videos can be quite elaborate, containing picture-in-picture frames and special animations or graphics. Such videos would take tremendous time and resources to develop. Videos could be as simple as a recording of one's lecture. There are also videos that fall somewhere in between, such as "whiteboard" style videos with voiceover audio. Instructors should consider the following when deciding on a format:

- How much material do you want each video to cover?
- What is the ideal length of time for your students to spend watching the videos?
- How much time and resources can you devote to editing and producing each video?
- Are the videos being developed for just your course, or will they be shared among instructors? If shared, will it be possible to collaborate such that each participant creates videos that can eventually be pooled?
- How will the videos be integrated with the pedagogy you plan to use in class?
- Are there existing videos that you can use for your course?

As an example, this author's 49 PreLectures averaged 19.3 minutes in duration and rarely exceeded 30 minutes. Yet, by choosing the whiteboard style of video and by using the time-saving strategy shown in figure 1, many of the videos are able to cover at least as much as would be covered in a typical class session. The PreLectures often include the typical examples and problems that organic chemistry instructors use to illustrate particular concepts. This allows the instructor to gauge student learning of the concepts with the first clicker question, and then proceed to more challenging problems if the majority of the class has demonstrated knowledge of the basics. Also, at the start of the project, the availability of instructional videos for organic chemistry was scarce, forcing this author to create his own. Today, there are ample instructional videos on the subject that one may choose to use (e.g., on the internet from other instructors or as part of textbook offerings from publishers). For instructors who may be hesitant to use the work of others, it could be possible to incorporate one's own perspective into the students' viewing experience by, for example, using a website or handout to provide text or audio commentary and instruction created by the instructor of the class.

Below are some points to consider when planning to flip one's course:

- Creating the videos:

    - Consider the positioning of the video recording device and microphones (if applicable).

- Will viewers be able to discern the images on the whiteboard?
- Will the actions of students in the audience (e.g. chatting with others, texting, browsing the internet) become distractions that detract from the learning experience?
- Will viewers be able to hear the instructor, as well as the questions and comments posed from the audience? The latter can be ameliorated by repeating all questions that students ask during class.

- Be deliberate in one's actions and the timing of one's comments:

  - If one plans to delete superfluous footage in order to shorten the duration of the video, take care not to speak during these moments or at least not to say anything that would be crucial to keep in the video.
  - Remember that there are two audiences, the students in the class and the students who will eventually be viewing the lecture as a video. One's delivery should cater to both.

- During production and editing:

  - Think about ways to supplement the video content. If one owns the video content, overdub mistakes or add subtitles. As suggested above, use strategies for incorporating the instructor's own perspective into the students' viewing experience.
  - Add interactivity or active learning into the videos.
  - Edit out unnecessary footage or audio.

- In the classroom:

  - Review the literature on active learning pedagogies to decide what will work best for the instructor and students.
  - Hold students accountable for viewing the videos prior to class time:

    - Make it a part of their grade.
    - Do not lecture in the same manner or on exactly the same topics that the videos cover. In other words, do not repeat the videos in class.

- Refer to specific parts of the videos when appropriate. Remind students of topics, problems, or techniques covered in the video.

- Reinforce the effectiveness of the pedagogy:

  - Point out when more depth of coverage has been achieved in class as a result of the flipped approach (e.g., when more difficult, exam-like problems are covered in class)
  - Remind students of the benefits of the flipped approach to their learning.
  - Include value-added activities that students will appreciate (e.g. sessions on obtaining summer research fellowships), and indicate that the flipped approach has made this possible.
  - Take opportunities to compare the new approach to prior approaches to teaching the class, especially when a benefit has been derived (e.g., a higher exam average than previous years).

- Start slow. Phase in the flipped approach by piloting certain chapters or topics.
- Assess the effectiveness of the approach by seeking feedback from students and from one's peers.
- Be willing to adapt to feedback and one's own evaluation of the approach.
- Consult or collaborate with those who have already flipped their courses.

Conclusion

This chapter has provided examples of the flipped classroom approach in teaching organic chemistry and strategies for adopting a flipped approach in general. Using the 2004 ACS organic chemistry exam to assess cumulative student learning in the course, we found that student learning was significantly improved using the flipped classroom approach described herein. The flipped approach also enhanced the learning environment in the course, reinvigorated the instructor's teaching, and has ultimately, through the materials and curriculum developed, created more opportunities for the instructor to improve upon his teaching and other pursuits. The approach is still in its infancy and there is certainly much innovation to be achieved.

Acknowledgments

The author acknowledges the financial support of the Camille and Henry Dreyfus Foundation through its Special Grant Program in the Chemical Sciences and the Claremont McKenna College Teaching Resource Center, as well as the Southern Regional Meeting of the American Chemical Society for the opportunity to present this paper at its 2014 fall regional conference.

References

1. Lage, M. J.; Platt, G. J.; Treglia, M. Inverting the classroom: A gateway to creating an inclusive learning environment. *J. Econ. Educ.* **2000**, *31*, 30–43.
2. Milman, N. The flipped classroom strategy: What is it and how can it be used? *Distance Learn.* **2012**, *9*, 85–87.
3. Fulton, K. P. 10 Reasons to Flip. *New Styles Instr.* **2012**, *94*, 20–24.
4. Christiansen, M. A. Inverted Teaching: Applying a New Pedagogy to a University Organic Chemistry Class. *J. Chem. Educ.* **2014**, *91*, 1845–1850.
5. Ealy, J. Development and Implementation of a First-Semester Hybrid Organic Chemistry Course: Yielding Advantages for Educators and Students. *J. Chem. Educ.* **2013**, *90*, 303–307.
6. Herreid, C. F.; Schiller, N. A. Case Studies and the Flipped Classroom. *J. Coll. Sci. Teach.* **2013**, *42*, 62–66.
7. Horn, M. B. The Transformational Potential of Flipped Classrooms. *Educ. Next* **2013**, *13*, 78–79.
8. Rosenberg, T. Turning Education Upside Down, *New York Times*, October 9, 2013. http://nyti.ms/1BzMaYM (accessed March 28, 2014).
9. Salemi, M. K. Targeting Teaching. An Illustrated Case for Active Learning. *South. Econ. J.* **2002**, *68*, 721–731.
10. Bonwell, C. C., Eison, J. A. *Active Learning: Creating Excitement in the Classroom*; ASHE-ERIC Higher Education Report No. 1; The George Washington University, School of Education and Human Development: Washington, DC, 1991.
11. Johnson, D. W., Johnson, R. T., Smith, K. A. *Active learning: Cooperation in the college classroom*; Interaction Book Company: Edina, MN, 1991.
12. Michael, J. Where's the evidence that active learning works? *Adv. Physiol. Educ.* **2006**, *159–167*, 2006.
13. Zappe, S., Leicht, R., Messner, J., Litzinger, T., Lee, H. W. "Flipping" the Classroom to Explore Active Learning in a Large Undergraduate Course. Paper presented at the American Society for Engineering Education Conference, Washington, D.C. 2009; AC 2009-9. http://goo.gl/aeCgl7 (accessed June 4, 2015).
14. Jensen, J. L.; Kummer, T. A.; Godoy, P. D. d. M. Improvements from a Flipped Classroom May Simply Be the Fruits of Active Learning. *CBE—Life Sci. Educ.* **2015**, *14*, 1–12.
15. Wison, S. G. The Flipped Class: A Method to Address the Challenges of an Undergraduate Statistics Course. *Teach. Psychol.* **2013**, *30*, 193–199.

16. Raths, D. 9 Video Tips for a Better Flipped Classroom. *Transform. Educ. Technol.* **2013**, *40*, 12–18.

17. Schwier, R., Misanchuk, E. R. *Interactive Multimedia Instruction*; Educational Technology Publications, Inc.: Englewood Cliffs, NJ, 1993.

18. Bacro, T.; Gilbertson, B.; Coultas, J. Web-delivery of anatomy video clips using a CD-ROM. *Anat. Rec.* **2000**, *261*, 78–82.

19. Nicholls, B. S. Pre-laboratory support using dedicated software. *Univ. Chem. Educ.* **1999**, *3*, 22–27.

20. Daniel, J. Making sense of MOOCs: Musings in a maze of myth, paradox and possibility. *J. Interact. Media Educ.* **2012**Art–18.

21. Conole, G. MOOCs as disruptive technologies: strategies for enhancing the learner experience and quality of MOOCs. *Rev. Educ. Distancia* **2013**, *39*, 1–17.

22. Emanuel, E. J. Online education: MOOCs taken by educated few. *Nature* **2013**, *503*, 342–342.

23. Thompson, C. How Khan Academy is changing the rules of education. *Wired Magazine* **2011**, *126*, 1–5.

24. Dewar, T.; Whittington, D. Online learners and their learning strategies. *J. Educ. Comp. Res.* **2000**, *23*, 385–404.

25. McBrien, J. L.; Cheng, R.; Jones, P. Virtual spaces: Employing a synchronous online classroom to facilitate student engagement in online learning. *Int. Rev. Res. Open Distrib. Learn.* **2009**, *10*, 1–17.

26. Song, L.; Singleton, E. S.; Hill, J. R.; Koh, M. H. Improving online learning: Student perceptions of useful and challenging characteristics. *Internet Higher Educ.* **2004**, *7*, 59–70.

27. Kearney, M.; Treagust, D. F.; Yeo, S.; Zadnik, M. G. Student and Teacher Perceptions of the Use of Multimedia Supported Predict–Observe–Explain Tasks to Probe Understanding. *Res. Sci. Educ.* **2001**, *31*, 589–615.

28. Singh, P.; Pan, W. Online education: Lessons for administrators and instructors. *Coll. Stud. J.* **2004**, *38*, 302–308.

29. Meyer, A., Rose, D. H., Gordon., D. *Universal Design for Learning: Theory and Practice*; CAST Professional Publishing: Wakefield, MA, 2014.

Technology

Chapter 4

E-Textbooks and the Digital Natives: A Study of First-Year Chemistry Students' Attitudes toward E-Textbooks

T. O. Salami[1,*] and E. O. Omiteru[2]

[1]Department of Chemistry, Valdosta State University,
1500 North Patterson Street, Valdosta, Georgia 31698, United States
[2]Center for Program Assessment, Analytics, and Evaluation,
Dewar College of Education and Human Services, Valdosta State University,
1500 North Patterson Street, Valdosta, Georgia 31698, United States
*E-mail: Tosalami@valdosta.edu

This study assesses digital natives' behavioral intentions to use e-textbooks in a freshman chemistry class. Our study combines four constructs from the Unified Theory of Acceptance and Use of Technology (UTAUT) and Technology Acceptance Model (TAM). Our results show that performance expectancy and effort expectancy are better predictors of students' intentions to use e-textbooks. An open-ended question reveals differences in students' viewpoint of the usefulness of e-textbook. This study also provides strategies which could make the use of e-textbooks more appealing to digital natives.

Introduction

Digital Natives

"Digital natives" is a term first introduced by Prensky (*1–4*) in early 2000. The term was used to describe and identify a generation of students born into technology who actively use technology and the Internet. Another term, the "Millennials", coined by Strauss & Howe (*5*) in the same time period, describes the same group of students.

The general assumption about the learning behavior of "digital natives" based on Prensky's articles (*1–4*) is that digital natives prefer visuals over text, function better in groups, have a short attention span, have digital technology skills, and need to be educated differently with the use of technology to help them learn better. Furthermore, Prensky (*1*) suggested that "it is very likely that our students' brains have physically changed and are different from ours as a result of how they grew up." This bold claim has not been substantiated, and several discussions are still ongoing.

Other schools of thought suggest otherwise. Kennedy and co-researchers (*6*) propose that the ability of students to use technology for entertainment (e.g., video games, movies, etc.) does not necessarily transfer to its use for educational purposes. Koutropoulos (*7*), in his article "Digital Natives: Ten Years After", suggests that several of the learning styles proposed for digital natives are not substantiated. Also, Thompson investigated the notion that "digital natives" think and learn differently due to their exposure to technology (*8*). His result shows a less deterministic relationship between technology and learning.

In light of this, as instructors, we often wonder if the use of technology is the major factor that enhances learning for "digital natives". We have observed variations and differences in students over the years, especially in their levels of comfort around technology. Our view is that students have different attitudes and motivation when it relates to technology. While some students enjoy using technology, some do not, and others are indifferent. Several factors such as pedagogy, type and deployment of technology in the classroom, quality of instruction and other intangibles also come into play.

What Are E-Textbooks?

An e-textbook is a book that is accessible digitally using a device that has an Internet connection (*9*). E-textbooks may be accessed electronically on computers, e-readers, PDAs, laptops, tablets and through mobile devices.

Previous studies have recognized that the convenience gained through the portability and a myriad of other features associated with e-textbooks has made this technology pervasive not only within society, but among educators (*10–12*). As more studies emerge on the advantages of e-textbooks and how they may benefit students, colleges have jumped at the possibilities of using them (*13–16*). Publishers have joined the mainstream and have started designing e-textbook packages that are marketed to students with the added promises of cost reduction and improved accessibility (*11*). Most students, especially those in higher education, are confident in their use of consumer technology (*17*). However, the

46

transition to using those technologies in higher learning still poses a challenge. The idea of using e-textbooks for anywhere and anytime access may seem like an excellent solution, but the magnitude to which this new method of learning will affect college students needs to be studied.

Several studies have compared the limitations and successes of e-textbook adoption in higher education (*18–20*). The conclusions in these studies indicate that users' preferences and perceptions about the use of e-books (*21*) and e-textbooks are mixed. While some studies report users' positive attitudes about their use, concerns remain about some "…real limitations based on the usability of e-book platforms …" (*12, 21*). A study by Nicholas and Lewis (*22*) asserted that the drawbacks in the use of e-textbooks by students might be attributed to concerns such as eyestrain from electronic displays, limited battery life for various devices, technology failures and other technical difficulties. Many of the design drawbacks of earlier e-readers and other devices used in accessing e-textbooks have been addressed by the emergence of technological advances in designs and features.

The interactive and dynamic nature of e-textbooks on most current devices can allow for tasks such as highlighting, rotating chemical structures, 3-dimensional view of structures, video links, and the ability to practice problems with built-in intuitiveness which offers students meaningful hints. This may also increase students' engagement in learning. Several chemical demonstrations may also be accessed through multimedia components associated with e-textbooks.

A recent study (*23, 24*) on the use of e-textbooks revealed that, despite these advantages, many students are yet to fully embrace e-textbooks. The biggest barrier to their use may be the culture of reading on paper rather than on the screen (*25–28*). The Book Industry Study Group (BISG) in August 2013 reported that despite the higher cost of traditional textbooks and everyday use of technology by college students, only 6% of college students use e-textbooks in their classes (*29*). Another factor that is contributing to the low use of e-textbooks by "digital natives" is the growth in the rental market (*30*). The rental market is cutting into the market share of traditional textbooks, and the cost of rental is comparable to the cost of e-textbooks, which gives more choices to the student. Watson's study (*31*) cautioned that, though the use of multimodal learning techniques is becoming more widespread, discussion around the pedagogical implementation in the classroom is still vital and should be ongoing.

Technology Use in Chemistry

The choice of textbook plays an important role in the design of a freshman chemistry course. Illustrations, graphics, enhanced text, and font arrangement in chemistry textbooks enhance reading and help with understanding of the content. Souza (*32*) studied the important role images and graphics play in chemistry textbooks. According to the study, images are important in chemistry because they are used to illustrate specific chemical concepts in chemistry textbooks. Some of the most common categories identified by Souza are structural models, experimentation, graphs, and diagrams, just to mention a few.

For example, a topic central to any introductory chemistry textbook is the concept of chemical bonding. To understand chemical bonding, instructors must focus on concepts such as charge, octet, and electron pairs in addition to ionic, covalent, and metallic bonding (33). Although many of these concepts are well illustrated by graphics in traditional textbooks, e-textbooks and companion websites have interactive graphics and models that may help students understand the bonding concept better. Nevertheless, incorporating technology into a chemistry class should not be done for the "wow factor" (34); it should be used to add an instructional opportunity and value. Lecture recordings (or podcasts), Short Message Service (SMS) polling, e-textbooks, and laboratory demo videos could also complement teaching.

The goal of any chemistry instructor is to make class time effective, encourage student participation, keep students engaged and reinforce problem-solving skills. Chemistry instructors must find a way to impact knowledge and concepts in the classroom by utilizing various methods. In the book "Methods of Teaching Chemistry" Forster (35) discusses several methods for teaching chemistry (See Table 1) and highlights the advantages and disadvantages of these methods. The list in Table 1 is not exhaustive; however, several of the aforementioned techniques serve as a foundation for newer methods of teaching chemistry.

Table 1. Summary of Methods for Teaching Chemistry. Adapted from reference (35). Copyright 2009 Global Media.

Method Types	Definition
Lecture method	The instructor talks and students are passive listeners
Lecture/ demonstration	Combines the instructional strategy of 'information imparting' and 'showing how"
Heuristic method	Discovery method of learning
Assignment method	Students given assignments to help problem solving skills
Project method	Activities related to the course content carried out in a natural setting
Unit method	Field activity where students are actively involved in learning process
Historical method	Learning the historical progression of a subject matter
Discussion method	Chemistry concepts are discussed in groups
Inductive method	Conclusion based on observation (e.g. color change in litmus paper)

Several chemistry professors are experimenting with the hybrid "flipped classroom" model (36, 37). The "flipped" method of teaching involves moving the "delivery" of course material from the traditional or formal class time to outside the formal class time through the use of videos, digital textbooks and other visuals, thereby allowing class time to be used for more interactive team activities related to topics being discussed/taught in class. Another creative method of teaching chemistry that may be combined with technology is the use of poetry writing, poster illustration, and group presentation (38). These methods might help instructors present chemistry to students as a fun and creative science. In order to augment learning, chemistry textbook publishers (39–41) have started implementing changes by packaging e-textbooks with "adaptive learning (data driven individualized learning) tools and technology".

Despite the availability and the multimedia features available in e-textbooks for the most popular general chemistry textbooks, the adoption rate of e-textbooks remains very low (29). Although the literature is replete with advantages and disadvantages of e-textbooks, it may be difficult to predict what the future holds for e-textbook usage in higher education. Moreover, mobile technology is a changing industry, where technologies are deployed on a continual basis; therefore, more research must be done to address conflicting views on the adoption and usage of e-textbooks by students in higher education. This study contributes to the ongoing conversation.

Research Model and Hypotheses

As a professor at a regional university in the southeastern part of the United States, I teach freshman chemistry among other courses. The students in the class (approximately 100–190 students each semester) are given the option to choose between an e-textbook and a traditional textbook. This information is clearly stated in the syllabus. Also included in the syllabus are the costs of the e-textbook and the traditional textbook. Because the e-textbook is almost half the price of the traditional textbook, the expectation is that most of the students, being "digital natives", will opt for the e-textbook. However, several surveys conducted prior to this study consistently show that many of the students (70–80%) are not using e-textbooks as expected (42).

To understand the underlying factors leading to the disparity in the choice of traditional textbooks over e-textbooks, an initial pilot study that examined students' perceptions about e-textbook usage in an introductory chemistry course was conducted. A review of the pilot study indicated that over 70% of the class used the traditional textbooks (42). The study concluded that students are reluctant to make the transition to e-textbooks for various reasons. Some of the reasons stated include probable technology failure, cost of paying a provider for internet access, and the inability of the students to purchase the e-textbooks using financial aid money. The issue of accessibility was crucial to students, especially after the book subscription comes to an end. This issue was important to students whose intentions were to use the e-textbook to study for entry tests into medical, veterinary, pharmacy or graduate schools. Nonetheless, students who purchased e-textbooks cited advantages such as portability and multimedia components as some of the reasons they elected to use it for the course.

This study answered some important questions, but left some unanswered. In particular, we wanted to know why students chose the e-textbook, because such information may translate to a better way of integrating e-textbooks into chemistry curriculum. Thus, our current study seeks to gain a better understanding of the reasons why some students chose the e-textbook.

This study utilizes constructs from the Unified Theory of Acceptance and Use of Technology (UTAUT) model developed by Venkatesh *et al.* (*43*). UTAUT identifies four constructs as direct determinants of user acceptance and usage behavior (Performance Expectancy, Effort Expectancy, Social Influence and Facilitating Conditions). The model, originally tested within the organizational setting, was established to be a useful tool for determining users' acceptance of new technologies. Other studies that used this theory to determine students' perceptions of technology found the theory to be an adequate predictor of users' behaviors and intentions to use technology (*44–47*).

Two constructs and one moderator from the Unified Theory of Acceptance and Use of Technology (UTAUT) model and two constructs from Technology Acceptance Model (TAM) (*48–53*) were used in our study to assess students' behavioral intentions to use e-textbooks. Altogether, these four constructs were tested against one moderator, Technology Experience, to determine users' intentions to use e-textbooks. (See Figure 1).

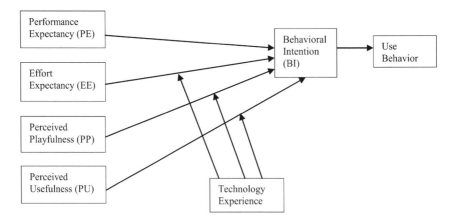

Figure 1. Research Model. (Adapted from references (43) and (49). Copyright 2003 and 1989 MIS Quarterly).

The two UTAUT constructs are Performance Expectancy (PE) and Effort Expectancy (EE). Performance Expectancy as defined by Venkatesh *et al.* (*43*) is the level at which an individual believes that the use of technology would help advance the person's job performance. According to Venkatesh, PE is a strong determinant of users' intention to use technology both in voluntary and mandatory work settings.

Effort Expectancy determines the extent of ease associated with the use of technology (43). This construct was measured three times in both voluntary and mandatory settings, and these studies found that the construct became less significant as usage increased (43).

The TAM constructs we used are Perceived Usefulness (PU) and Perceived Playfulness (PP). Perceived Usefulness is defined as "the degree of which a person believes that using a particular system would enhance his or her job performance" (52). In a study of users' intrinsic attitudes towards the World Wide Web acceptance, Moon and Kim (54) define Perceived Playfulness as "The extent to which the individual perceives that his or her attention is focused on the interaction with the world wide web". The validity of the "perceived usefulness" construct has been tested in previous studies in explaining users' attitudes and use behaviors towards technology (52). Our study assumes that, as "Digital Natives" who are technology savvy, the playful aspect of technology and the usefulness of the technology they are accustomed to (cell/mobile phones, tablets, iPods) would influence students' decisions in choosing e-textbooks. Furthermore, it is important to see if "digital natives" comfort and ease of use of entertainment technology translates to their choice of the e-textbook.

Survey questions were modeled after UTAUT and TAM technology acceptance models. Questions on the survey addressed the four key determinants and the moderating item. The survey used a 7-point Likert scale, with the highest number representing users who "strongly agree" with the question and the lowest number representing those who "strongly disagree." An open-ended question was added to the survey in this study to provide students with the opportunity to give more detailed information on the reasons why they opted for the e-textbook. The direct determinants (Performance Expectancy, Effort Expectancy, Perceived Playfulness and Perceive Usefulness) are investigated to determine how users' Behavioral Intentions will be influenced when moderated by technology experience.

The following hypotheses guided this study:

H1: Performance Expectancy (PE) will impact users' behavioral intentions to use e-textbooks

H2: Effort Expectancy (EE) influences behavioral intentions when moderated by Technology Experience

H3: Perceived Playfulness (PP) influences behavioral intentions when moderated by Technology Experience

H4: Perceived Usefulness (PU) will influences behavioral intentions when moderated by Technology Experience

H5: Behavioral Intentions (BI) will have an impact on e-textbook usage for students

Methodology

Two of the objectives of this study are to examine students' experiences with e-textbook usage in the classroom and to determine its efficiencies from the students' perspective. This study adopts a non-random purposeful sampling technique. This technique gathers data strategically, depending on the study purpose and resources (55). Out of 150 undergraduate students who enrolled in the chemistry class, only forty-six students who purchased and used e-textbooks participated in this study. Overall, twenty-six male and twenty female students completed the survey.

Results and Discussion

Data Reliability

Cronbach's alpha was calculated for each construct to determine the internal consistency of the Likert-response items that were used on the survey for e-textbook users. According to George *et al.* and Matkar (56, 57), a value of 0.70 or higher is considered acceptable, and it indicates data reliability and consistency. (See Table 2). The PE and EE values suggested a relatively high internal consistency and PU a moderate consistency.

Table 2. Cronbach's Reliability for Constructs

Construct	Number of items	Cronbach's Alpha
PE	4	0.867
EE	4	0.874
PP	4	0.448
PU	4	0.774

After establishing that there was no correlation between the independent variables, (and data was normally distributed), a stepwise multiple regression analysis was conducted to determine whether PE, EE, PP and PU could influence BI. Stepwise linear regression analysis was used in analyzing the data. The regression analysis showed a weak correlation for PU and PP ($p > 0.05$). PE and EE were found to be significantly related to BI at $F (2, 41) = 31.47$, $p < 0.001$, $R^2 = 0.79$, adjusted $R^2 = 0.59$ (See Table 3). The multiple correlation coefficient (R) was 0.78, indicating approximately 60.6% of the variance of BI could be accounted for in this model.

From these results, PE and EE are better predictors of intentions to use e-textbooks. Consistent with prior research (*43*), PE is the strongest predictor of students' intentions to use e-textbooks. Although PE (β = 0.57, p < 0.001) and EE (β = 0.29, p = 0.024) are strong predictors, the relative strength of PE is stronger (than EE) in predicting behavioral intentions to use e-textbooks. This supports our hypothesis *H1* (Performance Expectancy will impact users' behavioral intentions to use e-textbooks) and *H2* (Effort Expectancy influences behavioral intentions when moderated by Technology Experience). The beta value is a measurement of how strongly each predictor influences the dependent variable (Behavioral Intention). The high positive beta value indicates a strong influence (of PE and EE) on the dependent variable (BI). Model 2 (Table 3) of the stepwise regression analysis shows the beta coefficient for PE to be 0.57, therefore each unit increase of PE will trigger an increase in BI by 0.57 units. Our study cannot explain the influence of PP and PU on students' intentions and use behavior of e-textbooks *(H3)(H4)*. Although, Moon and Kim (*54*) found PP and PU to be significant in user attitudes, the influence of these determinants depend on whether the technology is used for fun or for studying (*58*).

Table 3. Regression Analysis of Data-Model Summary

Model	R	R Squared	Adjusted R Squared	Std. Error of the Estimate
1	0.743[a]	0.552	0.542	0.936
2	0.778[b]	0.606	0.586	0.890

[a] Predictors: (Constant), PE.   [b] Predictors: (Constant), PE, EE.   [c] Dependent Variable: BI.

Open-Ended Questions Response

The students' responses to the open-ended question are mixed. Some were very positive and excited about using the e-textbook while others were not. Selected comments are listed in Table 4. Several students indicated that the cost of e-textbook was the major reason for using it, *"E-textbooks are cheaper, and that is a big deal."* Other students emphasized the portability and the availability of multimedia functionalities, for example, the search and highlighting capabilities. According to one of the students, *"It's just a cheaper way to get a textbook, not necessarily better or worse, just cheaper"*. This comment is not in support of or against e-textbooks. Other deterrents mentioned in students' narratives include anxieties about technology failure and distractions from the multimedia features associated with e-textbooks. One student responded, *"I feel as if they are distracting and do not enhance learning"*.

Out of the 150 students registered for the class, only 46 (30.6%) used e-textbooks. It was also surprising that 22% of the 46 students we studied still expressed concerns about e-textbooks.

Table 4. Selected Comments from Open-Ended Questions

Selected Comments
• "I believe that e-textbooks are good because they are less expensive than regular textbooks"
• "E-textbooks are cheaper, and that is a big deal and it offers a search feature that can save time also I can read in the dark because the screen is lit and doesn't keep my roommate up"
• "Very convenient for busy students"
• "E-textbooks really helped me to understand my all of the class lectures"
• "They are easy to use"
• "They are accessed on the web portable…. if you live far away you cannot forget it you don't have to make the trip back to get your book"
• "It's just a cheaper way to get a textbook, not necessarily better or worse, just cheaper"
• "I feel as if they are distracting and do not enhance learning"

However, our expectation is that if we are able to understand the reasoning and the rationale guiding the choice of e-textbook then it is conceivable that we might begin to find better ways of making e-textbooks more appealing to students.

Perceived Playfulness and Usefulness

Our results showed that Perceive Playfulness (PP) and Perceived Usefulness (PU) are not significant predictors of students' intentions to use e-textbooks. This indicates that even though "digital natives" are comfortable with technology, this feeling is not replicated once it is a book used in learning. From the demography of students who participated in the study, 37% claimed to have excellent technological experience, 50% have good technological experience, and 13% of the students studied had a satisfactory technological experience. The students surveyed all claimed to have some sort of prior technological experience. This data are consonant with our conclusion that "digital natives" are comfortable with technology; however this interest is not significantly reflected in their choice of texts for learning.

Limitations

This study focused on students taking entry-level chemistry classes, and was limited to a small group of students (those who used e-textbooks). These students were from different departments in different stages of their college career. A more precise determination would have required that this study categorize students based on unique criteria such as the number of years in college, majors, gender or the device students chose for accessing their course readings. Future research should survey students based on these criteria to obtain an accurate representation. A sample size of 46 represents a limitation in this research.

Another limiting factor that was not investigated in this study was personal and institutional financial constraints. These findings may not be replicable in schools that make provisions of hardware and software for students to access e-textbooks. Furthermore, opinions are subjective and might have been influenced by other factors that were not addressed on the UTAUT and TAM survey questions and in the focus group.

Recommendations and Conclusions

The objective of our study was to identify the reasons why students purchased and used e-textbook for this class. A basic understanding of the challenges and advantages offered by e-textbooks provides useful insight for future integration of e-textbooks in the chemistry class or curriculum. Although our findings show that some students are willing to use e-textbooks, we anticipate that if the quality and potential of e-textbooks are clearly stated, more students may be willing to use the technology. Ultimately, if any educator decides to use e-textbooks, the goal of educators and stakeholders should be to help students have a smooth transition from a traditional textbook to e-textbooks. The educator should adhere to best practices, setting realistic expectations of e-textbooks and highlighting advantages such as reduced cost, environmental friendliness, and the ability to integrate keyword searches and multimedia features.

Several reports show that various universities and colleges are trying different approaches and are experimenting with e-textbook programs. The Virginia State University (VSU) did an exploratory study of the first year of a pilot program. In this program, 991 students in nine core courses at VSU Reginald F. Lewis College of Business (RFLCB) replaced traditional textbooks with openly licensed e-textbooks through Flat World Knowledge (59). A similar program was conducted at the University of Phoenix. This college combined designated textbooks for all courses into an electronic library and charged students $75 a semester to access textbooks (60). Another approach used by the University of Idaho professors provides e-textbooks with content tailored to specific courses and charges students course fees (61).

The increased interest from various universities in e-textbooks is prompting publishers to improve their products to enhance students' learning experience. McGraw Hill Education, in a report (39) entitled "Brave New World of Education: Personalized Adaptive Learning Tools Promise One-on-One Tutoring for All Students," explains that through their adaptive learning programs such as LearnSmart (an interactive study tool) and ALEKS (a web-based assessment and learning system) students can improve their learning experience. Pearson (40) and Cengage (41) have teamed up with Knewton, an adaptive learning platform which personalizes course material. Several chemistry departments have adopted many of these programs. The publishers have provided several studies to exhibit success of their products. However, a report by the Educational Growth Advisors (62) suggests that adaptive technology "may well conflict with the prevailing teaching paradigm at a given institution." Modest student outcomes, due to poorly prepared and executed implementations, could deter skeptical faculty from further exploring such technology.

Ultimately any instructor that chooses to use e-textbooks for a course must carefully consider how to fully incorporate this technology into the course content, thereby making the use of the e-textbook engaging for the students. This may involve writing additional notes, highlighting important facts or creating links to videos or games. The games could be a simple crossword puzzle about definitions, compounds, elements or nomenclature in chemistry.

Science educators need to be more involved and focus on ways to bring educational media under certain guidelines or benchmarks in the area of design, especially in the move towards free open-text books. Lastly, it is also central that Chemistry faculty are continually trained in various educational software and teaching methods. Instructors should also stay current on high-impact educational practices (*63*).

References

1. Prensky, M. Digital natives, digital immigrants. *On the Horizon.* **2001**, *9*, 1–6.
2. Prensky, M. Digital natives, digital immigrants: do they really think different? *On the Horizon.* **2001**, *9*, 1–6.
3. Prensky, M. Overcoming Educators' Digital Immigrant Accents: A Rebuttal. *The Technology Source*, 2003. http://technologysource.org/article/overcoming_educators_digital_immigrant_accent/ (accessed January 25, 2015).
4. Heppell, S. Foreword. In *Teaching Digital Natives: Partnering for Real Learning*; Prensky, M., Author; Corwin: Thousand Oaks, CA, 2010.
5. Howe, N., Strauss, W. *Millennials Rising: The Next Great Generation*; Vintage: New York, 2000.
6. Kennedy, G.; Judd, T. S.; Churchward, A.; Gray, K. First years students' experiences with technology: Are they really digital natives? *Aust. J. Educ. Tech. (AJET).* **2008**, *24*, 108–122.
7. Koutropoulos, A. Digital Natives: Ten Years After. *MERLOT Journal of Online Learning and Teaching*, **2011**, *7*, 525–538. http://jolt.merlot.org/vol7no4/koutropoulos_1211.htm (accessed January 25, 2015).
8. Thompson, P. The digital natives as learners: technology use patterns and approaches to learning. *Computers & Education [serial online].* **2013**, *65*, 12–33 [Available from: Science Direct, Ipswich, MA (accessed January 25, 2015)].
9. Margaryan, A.; Littlejohn, A.; Vojt, G. Are digital natives a myth or reality? University students' use of digital technologies. *Comput. Educ.* **2011**, *56*, 429–440.
10. El-Hussein, M., Cronje, J. C. Defining mobile learning in the higher education landscape. *J. Educ. Techno. Soc.* **2010**, *13*, 12–21. http://web.ebscohost.com/ehost/pdfviewer/pdfviewer (accessed January 26, 2015).
11. Butler, D. Technology: The textbook of the future. *Nature* **2009**, *458*, 568–570.

12. Wilkes, J.; Gurney, L. J. Perceptions and applications of information literacy by first year applied science students. *Aust. Acad. Res. Libr.* **2009**, *40*, 159–171.

13. Clough, G. G.; Jones, A. C.; McAndrew, P. P.; Scanlon, E. E. Informal learning with PDAs and smart phones. *J. Comput. Assist. Learn.* **2008**, *24*, 359–371.

14. Hlodan, O. Mobile learning anytime, anywhere. *Bioscience* **2010**, *60*, 682–682.

15. Shohel, M. C.; Power, T. Introducing mobile technology for enhancing teaching and learning in Bangladesh: teacher perspectives. *Open Learning* **2010**, *25*, 201–215.

16. Shen, J. The e-book lifestyle: An academic library perspective. *Reference Librarian* **2011**, *52*, 181–189.

17. Wilkes, J.; Gurney, L. J. Perceptions and applications of information literacy by first year applied science students. *Aust. Acad. Res. Libr.* **2009**, *40*, 159–171.

18. Tees, T. E-readers in academic libraries - a literature review. *Aust. Libr. J.* **2010**, *59*, 180–186.

19. Chong, P. F.; Lim, Y. P.; Ling, S. W. On the design preferences for ebooks. *IETE Technical Review* **2009**, *26*, 213–222.

20. James, P. *Mobile-Learning: Thai HE Students Perceptions and Potential Technological Impacts* **2011**, *4*, DOI: 10.5539/ies.v4n2p182.

21. In this work e-books will be considered as novels or any other type of electronic book for leisure reading. E-textbooks will be classified as books used for instructions for a specific class or course.

22. Nicholas, A. J.;Lewis, J. K. The Net Generation and E-Textbooks. eScholar@Salve Regina. http://escholar.salve.edu/fac_staff_pub/17. 2009. (accessed January 27, 2015).

23. Walton, E. W. From the ACRL 13th national conference: e-book use versus users' perspective. *College & Undergraduate Libraries.* **2007**, *14*, 19–35.

24. Shen, J. The e-book lifestyle: An academic library perspective. *Ref. Lib.* **2011**, *52*, 181–189.

25. Vernon, R. F. Teaching notes: paper or pixels? An inquiry into how students adapt to online textbooks. *J. Soc. Work. Educ. (JSWE)* **2006**, *42*, 417–427.

26. Nelson, M. R. E-books in higher education: nearing the end of the era of hype? *EDUCAUSE Review* **2008**, *43*, 40–56.

27. Shepperd, J. A.; Grace, J. L.; Koch, E. J. Evaluating the electronic textbook: is it time to dispense with the paper text? *Teach. Psychol.* **2008**, *35*, 2–5.

28. Nortcliffe, A.; Middleton, A. Smartphone feedback: using an iPhone to improve the distribution of audio feedback. *Int. J. Elec. Eng. Edu.* **2011**, *48*, 280–293.

29. Student Attitudes toward Content in Higher Education, 2013. Book Industry Study Group. http://www.bisg.org/publications/product.php?p=22 (accessed January 26, 2015).

30. Rosen, J. Beyond Book Rental: The Next Big Thing on Campus. *Publishers Weekly [serial online].* **2013**, *260*, 8–9 [(Literary Reference Center, Ipswich, MA (accessed February 5, 2015)].

31. Watson, J.; Puccini, L. Digital natives and digital media in the college classroom: assignment design and impacts on student learning. *Educ. Media. Int.* **2011**, *48*, 307–332.

32. Souza, K. A. F. D.; Porto, P. A. Chemistry and chemical education through text and image: analysis of twentieth century textbooks used in brazilian context. *Sci Educ.* **2012**, *21*, 705–727.

33. Croft, M.; de Berg, K. From common sense concepts to scientifically conditioned concepts of chemical bonding: an historical and textbook approach designed to address learning and teaching issues at the secondary school level. *Sci Educ.* **2014**, *23*, 1733–1761.

34. Harrison, C. R. The use of digital technology in the class and laboratory. *Anal. Bioanal. Electrochem.* **2013**, *405*, 9609–9614.

35. Forster, S. *Methods of Teaching Chemistry [e-book]*; Global Media: Chandni Chowk, Delhi, 2009, pp 71–103.

36. Hank, T. Blended and Flipped Learning. *Technology & Learning* **2013**, *34*, 44–48.

37. Scott, L. Flipped Learning. *Education Digest* **2013**, *79*, 13–18.

38. Furlan, P. Y.; Kitson, H.; Andes, C. Chemistry, poetry, and artistic illustration: an interdisciplinary approach to teaching and promoting chemistry. *J. Chem. Ed.* **2007**, *84*, 1625–1630(accessed February 14, 2015).

39. Macmillan White Paper. http://www.siia.net/visionk20/files/Brave% 20New%20World%20of%20Education.pdf (accessed February 14, 2015).

40. Information from Pearson site. http://www.pearsonmylabandmastering.com/ northamerica/masteringchemistry/students/titles-available/index.php (accessed February 14, 2015).

41. Information from Cengage site. http://owl.cengage.com/partners/ brookscole/epin.html (accessed February 14, 2015).

42. Omiteru, E. *Mobile Learning: An Evaluation of College Students' Perceptions and Use of Electronic Textbooks*; Valdosta State University: Valdosta, GA, 2013.

43. Venkatesh, V.; Morris, M.; Davis, G.; Davis, F. User acceptance of information technology: toward a unified view. *MIS Quarterly.* **2003**, *27*, 425–478.

44. Wang, H.; Wang, S. User acceptance of mobile internet based on the unified theory of acceptance and use of technology: investigating the determinants and gender differences. *Soc. Behav. Pers.* **2010**, *38*, 415–426.

45. Willis, M.; El-Gayar, O. F.; Bennett, D. Examining healthcare professionals' acceptance of electronic medical records using UTAUT. *Issues in Informational Systems.* **2008**, *9*, 396–401.

46. Birch, A.; Irvine, V. Pre-service teachers' acceptance of ICT integration in the classroom: Applying the UTAUT model. *Educ. Media. Int.* **2009**, *46*, 295–315.

47. Wang, Y.; Wu, M.; Wang, H. Investigating the determinants and age and gender differences in the acceptance of mobile learning. *Br. J. Educ. Tech.* **2009**, *40*, 92–118.

48. Davis, F. *A Technology Acceptance Model for Empirically Testing New End User Information Systems: Theory and Results*; Unpublished Doctoral Dissertation, MIT Sloan School of Management: Cambridge, MA, 1985.

49. Davis, F. Perceived usefulness, perceived ease of use, and user acceptance of information technology. *MIS Quarterly.* **1989**, *13*, 319–340.

50. Davis, F. User acceptance of computer technology: system characteristics, user perceptions. *Int. J. Man Machine Studies.* **1993**, *38*, 475–487.

51. Davis, F.; Venkatesh, V. A critical assessment of potential measurement biases in the technology acceptance model: three experiments. *Int. J. Human Computer Studies.* **1996**, *45*, 19–45.

52. Davis, F.; Bagozzi, R. P.; Warshaw, P. R. User acceptance of computer technology: a comparison of two theoretical models. *Manag. Sci.* **1989**, *35*, 982–1003.

53. Davis, F.; Bagozzi, R.; Warshaw, P. Extrinsic and intrinsic motivation to use computers in the workplace. *J. Appl. Soc. Psychol.* **1992**, *22*, 111–132.

54. Moon, J. W.; Kim, Y. G. Extending the TAM for a world-wide-web context. *Inform. Manag.* **2001**, *38*, 217–230.

55. Patton, M. Q. *Qualitative Research & Evaluation Methods*; Sage Publications, Inc.: Thousand Oaks, CA, 2002.

56. George, D.; Mallery, P. *SPSS for Windows Step by Step: A Simple Guide and Reference, 11.0 update*, 4th ed.; Allyn & Bac: Boston, MA, 2003.

57. Matkar, A. Cronbach's alpha reliability coefficient for standard of customer service in Maharashtra state cooperative bank. *IJBM* **2012**, *11*, 89–95.

58. Atkinson, M.; Kydd, C. Individual characteristics associated with World-Wide-Web use: an empirical study of playfulness and motivation. *Database Adv. Inform. Syst.* **1997**, *28*, 53–62.

59. Feldstein, A. P.; Maruri, M. M. *Int. Res. Educ.* **2013**, *1*, 194–201 (ISSN 2327-5499).

60. Blumensty, K. G. To cut costs, ought colleges to look to for-profit models? *Chron. High. Educ.* **2008** June 13, A19–20.

61. Baker-Eveleth, L.; Miller, J. R.; Tucker, L. Lowering business education cost with a custom professor-written online text. *J. Educ. Bus.* **2011**, *86*, 248–252.

62. Educational Group Advisors Report. http://tytonpartners.com/ (accessed February 14, 2015).

63. Kuh, G. D. High-impact educational practices: what they are, who has access to them, and why they matter. *AAC&U* **2008**.

Chapter 5

Clustered Discussion Board as an Online Tool To Enhance Student Learning and Participation

Jomy Samuel*

DeVry University, College of Health Science and College of Liberal Arts and
Science, Atlanta, Georgia, 30342, United States
*E-mail: jsamuel@devry.edu

Discussion board is a common online tool used in blended
and online courses. Peer-to-peer written interactions through
online platforms like discussion board are known to enhance
learner understanding and contribute positively to the learning
environment. Although research has shown that discussion
boards can greatly enhance chemistry education, it is not widely
used in chemistry curricula. The traditional online discussion
board starts with one initial post based on the discussion topic,
and the subsequent posts contribute additional content that
extends the discussion horizontally. Science courses are fact
based and hence not amenable to long horizontal discussions.
Furthermore, information gets lost in the hundreds of posts
and threads. Here I have presented a new framework for
online discussion board based on "clustered discussions" as
an alternative to the traditional "horizontal discussions". The
clustered discussion board framework segregates the discussion
posts based on pre-defined criteria making it more effective in
encouraging student participation and student learning while
being less taxing on the instructor.

Introduction

Over the past decade, hybrid or blended courses have been on the rise both in K-12 and higher education (*1–3*). Studies have shown that students like the convenience and format of a blended program (*4–6*). However, engaging the students in active learning online can be a challenge.

Asynchronous discussion board is a widely accepted tool used for interactions in the online classroom (*7, 8*). One of the advantages of an online discussion board is that it allows peer-to-peer written interaction that is known to contribute positively to the learning environment, motivate students, and keep them engaged in the course (*7, 9–12*). Talking and writing about science topics helps students develop a deeper understanding of concepts and helps them hypothesize, clarify and distribute knowledge among peers (*9, 13, 14*).

Although research has shown that discussion board can greatly enhance chemistry education, it is not widely used by faculty in chemistry curricula (*15*). The possible reason for this is that chemistry is one of the basic sciences and not exactly tailored for online or offline lengthy discussions in the traditional sense. This is true for other science courses too. This chapter explores the benefits and limitations of a traditional horizontal online discussion board and presents a new framework based on "clustered discussions" that is more effective in encouraging student participation and student learning while being less taxing on the instructor.

Traditional "Horizontal" Online Discussion Board

Online asynchronous discussion board is widely used as part of the course curriculum in a wide range of disciplines. The broad acceptance of this online tool can be attributed to its familiarity and ease of use. When designing an online discussion board, the most common approach is to take a topic that instructors would normally present to students in a face-to-face setting and transfer the exercise online. The goal is to take advantage of peer-to-peer interaction and learning without having to devote precious class time that could be used for more instructor-student interaction.

A successful traditional online discussion starts with an initiating post based on the discussion topic. The discussion then moves horizontally with the participants expressing their analyses, opinions and feedback on the original topic and contributing additional content (Figure 1).

Advantages of Traditional Online Discussion Board

Besides saving class room time, an online discussion board has the following advantages:

- It supplements instructor-prepared course material with student-generated alternative perspective.
- It allows the student time to research the topic, analyze it and then respond at a time that is conveneint to them without experiencing the pressure they feel in a classroom setting.

- It personalizes the learning experince. Each student is able to reflect upon the concept and understand it in his/her own way at their own pace.
- All students in a class participate in the discussion, which is not possible in a classroom face-to-face discussion.
- It is possible to have multiple communications (threads) conducted simultaneously.

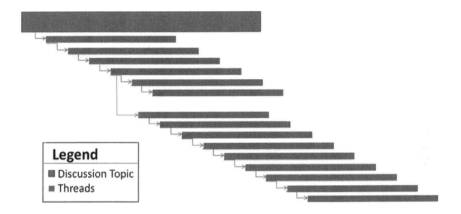

Figure 1. Schematic representation of a "Traditional Discussion Board".

Limitations of the Traditional Online Discussion Board

In subjective liberal arts and business courses, there are multiple viewpoints on every topic that the students may discuss. These topics easily lend themselves to long horizontal threads in an online discussion board. Science courses are based on facts, not viewpoints, and hence do not lend themselves to long horizontal discussion threads. For example, consider an online asynchronous discussion on chemical formulas and naming of molecules. For a traditional online discussion board, the discussion topic will be framed as a general question (Table 1). When you have a class of 24–34 students, it is hard find something new to say about "how to name a molecule" in 24 different ways! So the students will try to just be creative with words instead of actually learning how to name a molecule. This lack of concrete learning outcomes discourages students and the instructor.

Table 1. Traditional Discussion Topic: Naming a Molecule

Topic Description:
'Let's begin this discussion by talking about the naming of different compounds. Why would it be important to make sure that you know which compound is associated with which name? How would you go about naming a molecule?'.

Beyond the limitations specific to science courses, one of the main challenges of a traditional discussion board is the sheer number of threads and posts within each discussion and their lack of organization. It often requires students to click through hundreds of posts within multiple unorganized, disjointed threads to follow the discussion and to find relevant information (*16, 17*). In the topic given in Table 1, to keep the discussion moving, students could use examples of specific molecules to explain their answers, and the instructor could ask follow up questions. However, these examples, explanations and questions get lost in the threads and posts. This discourages student engagement.

The Re-Designed "Clustered" Online Discussion Board

Clarity of discussion topic as well as the purpose and outcome of the learning activity is of paramount importance to student participation and engagement in an online discussion board (*18*). Clustering of posts in different threads of a traditional "horizontal" online discussion board has been used in algorithms for various studies (*16, 17*). However, these clustering operations are used as an analytical tool to extract useful information from the repository of data contained within the posts. Since the clustering of posts is done after the end of discussion, it does not help the students in their engagement or learning goals.

In the proposed new framework for the online discussion board, the discussion posts are automatically clustered based on a pre-defined criteria (Figure 2). In this new clustered online discussion board framework, every student is required to kick off a new discussion with a sub-topic (vertical movement). The horizontal movement of the discussion threads start from these sub-topics and progresses with students adding new content and interracting with the goal of helping each other achieve their learning goals. All these conversations, although independent, have the main discussion topic as their common element. Thus, in a class of 30 students, there will be 30 sub-topics and as many simultaneous discussions ongoing at any given time. This framework inherently overcomes the challenge of data mining in the traditional discussion board. Students can easily find and follow discussions, which keeps them engaged.

The components of the clustered online discussion board are the same as that of a traditional discussion board—the discussion topic/question, instructions for the students, facilitation and feedback by the instructor and the grading rubrics. Out of these four components, the discussion topic and the accompanying instructions dictate the framework of a discussion board and the tone and success of any discussion.

An example of a discussion topic framed for a traditional discussion board was given in Table 1. As explained in that section, the subject of chemical formula and naming of molecules does not lend itself to a productive traditional "horizontal" discussion. In the following sections, I will illustrate how the same discussion topic and instructions could be re-framed for a clustered online discussion board.

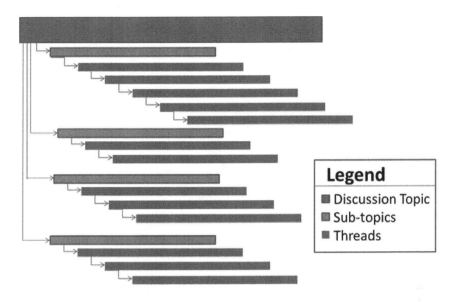

Figure 2. Schematic representation of a "Clustered discussion board".

Instructions

The success of any online discussion board and the quality of posts by each student depends on the clarity of the discussion topic and the instructions accompanying it. Proper instructions are crucial in supporting both low-level cognitive behaviors such as knowledge sharing as well as higher-level exploration with the integration of ideas. If aligned with grading rubrics, these instructions can also serve as extrinsic motivation for participation (*19*). Further, the role of the instructor and students must be clearly articulated from the start and throughout the learning activity.

For best results in a clustered discussion board, instructions should be provided at two levels:

1. **General instructions** (Table 2) given at the start of the term lays the foundation for a successful online discussion throughout the term. This should specify grading criteria and the frequency and type of posts expected.
2. **Topic specific or "What to post" instructions** for each discussion topic should specify the subject line for their initial sub-topic post, expected content in these posts and some general guidelines on response posts.

Table 2. General Instructions for a Clustered Discussion Board

Each week we will be doing an online discussion based on the topic covered in class. The discussion topic is designed with the goal of helping you understand the concepts, research and apply the concepts and practice problems. Before posting on the discussion board, please read the topic and instructions very carefully.

- For each discussion topic, you should have a minimum of 3 total posts-
  ○ 1 initial "sub-topic" post in direct response to the discussion topic/question. This post will have a new subject line.
  ○ A minimum of 2 responses to other students' posts
- Posts will be graded on Frequency (dates that you post) and Quality of each post.
- Each week you are required to do a minimum of one (1) post on three (3) different days.
- Each week the discussion starts on Sunday midnight and ends at 11:59 pm (EST) on Saturday
- Your first post needs to be in by Wednesday to receive full points.

Grading criteria:

- Frequency of posts (minimum one post on three different days) - 45%
- Quality of posts:
  ○ Initial sub-topic post - 25%
  ○ Response posts (minimum 2) – 20%
- First post by Wednesday – 10%

Initial "sub-topic" post:

The sub-topic post directly addresses the discussion topic. To do the sub-topic post, make sure you read the discussion topic carefully. At the start of each week, you will see a post from the instructor titled "What to post" that gives you more specific instructions on what to write in the subject line and content of your sub-topic post. The discussion topic is designed in such a way that each student will have a unique subject line. Make sure you follow the instructions for the subject line and the content. Posts that do not follow the instructions given for the discussion topic and under "What to post" will not receive full "Quality" points.

Response post:

Each student is required to do at least 2 response posts each week. You are encouraged to do more. When you respond to a post by another student, please address the person you are responding to. Be courteous, polite and frame your sentence correctly. A graded response post should add some new information to the sub-topic post. A response post that simply conveys "Thank you" or "Great job" or "I agree" will not be counted towards graded responses. Having said that, if someone helps you make a correction to your initial post or helps you understand a concept, you are encouraged to thank them. Your etiquette on the discussion board will be considered towards the overall "Quality" of your posts.

Providing the grading criteria at the start of the term helps students improve their posts. For the first week of online discussion, it also helps to go over these instructions in class. In my experience, it does take the students one or two weeks of discussion postings to fully understand the new discussion framework. However, after about two weeks they start seeing the benefits of it and are motivated to participate in the discussions. In the next section, when I explain the framing of a discussion topic in a clustered discussion board, I will provide more details and examples of the topic specific "What to post" instructions.

Discussion Topic

The main difference between a traditional "horizontal" discussion board and a clustered discussion board is the way the discussion topic is framed. In a traditional discussion board, one initial post based on the topic kicks off the discussion that moves horizontally through the threads. However, in a clustered discussion, the discussion topic provides a framework around which multiple simultaneous discussions happen. In other words, in a clustered discussion board, the discussion topic should have a "multiplier effect". Table 3 gives an example of a discussion topic on the subject of chemical formula and naming of molecules framed for a clustered discussion board.

Table 3. Clustered Discussion Topic – Naming of Compounds

Topic Description:
Pick any compound that contains a polyatomic ion and give the steps you would follow to arrive at its chemical name. Each student is required to make sure that the compound they post has not already been done by another student. If you post a compound that has already been done by another student, your sub-topic post will not receive any grades. So please read the posts by your classmates before writing your post.

In the above example (Table 3), the discussion topic provides the guidelines for each student to pick a compound and start their own discussion sub-topics. This propagates the vertical movement of the discussion. This multiplier effect is missing in the discussion topic for a traditional discussion board given in Table 1.

In order to pick a compound for the post, the student needs to first know what a polyatomic ion is. Then they have to read their textbook or search on the internet to find the chemical formula of a compound containing a polyatomic ion. Furthermore, the topic description clearly states that each student has to pick a compound that has not already been posted by another student. This motivates the student to get started on the discussion board early so that they have a wider range of possible compounds. It also requires them to look through the sub-topics posted by other students.

The topic specific "what to post" instructions (Table 4) give students clear instructions on what they are expected to post on the subject line and content of thier initial sub-topic post. These instructions are usually provided as the first post by the instructor to kick off the discussion. However, it can also be included right below the topic description. Having a separate subject line for each compound (sub-topic) automatically categorizes the posts and the subsequent threads into well defined clusters of discussions within the discussion board. Students and the instructor can easily scroll through all the sub-discussions going on within the class, filter content and pick the discussion that they would like to read and participate in. This also allows the instructor to do follow-up questions as a sub-topic post that is easily visible to all students.

Table 4. Example of Topic Specific "What To Post" Instructions

<u>**"What to post" Instructions:**</u>
Your initial sub-topic post should be done as a direct response to the discussion topic. It should have a new subject line.
• <u>Subject line</u>: Please put the chemical formula of the compound in the subject line.
• <u>Content</u>: Please provide the following information in your sub-topic post:
○ Cation name and symbol with charge
○ Anion name and symbol with charge
○ Type(s) of bonds (explain)
○ Give the chemical name of the compound
<u>**Examples of possible graded responses:**</u>
• If a student has made a mistake in their sub-topic post or one of their responses, you are encouraged to politely point it out, explain the concept and give them hints for correcting their mistake. Please do not give the answer. You should be helping your classmates learn how to name a compound, not just do it for them.
• Calculate the molar mass for the compound being discussed.
• Calculate the number of grams in 0.38 moles of the compound being discussed

Furthermore, providing clear instructions on what is expected within the content of the initial sub-topic post helps the student formulate quality discussion posts and achieve specific learning goals. In the example presented here, for each compound (sub-topic), the student will need to be able to pick apart the cation and anion of the compoud, figure out the type(s) of bond in the compound and then name the compound. This exercise forces them to gain deeper understanding about ions, compounds, types of bonds, chemical formulas and naming of compounds.

According to the general instructions (Table 2) for the discussion board, each student is required to write a minimum of two response posts. So using the instructions and grading rubrics, the instructor is able to create not just opportunities for students to interact, but the requirement that they do so. The "possible graded responses" instruction (Table 4) guides the student in their responses. In the example given here, in order to do the response post, the student will have to not only read through the posts, but will also need to work through the steps for naming of the compounds posted by their classmates. This exercise forces them to practice, research, analyze and gain a deeper understanding of the concept. When responding to the incorrect posts and explaining how they arrived at their conclusions, the student is forced to formulate an explanation in their own words. This further reinforces their reasoning ability and teaches them how to communicate and collaborate effectively. Using the "possible graded responses" guidelines the instructor is also able to connect the concept of chemical formula and naming to the concept of molar mass and stoichiometry. By giving the questions on molar mass and mole as an optional response post, the instructor is allowing the students to guide the pace of the discussion and learning.

Tables 5 and 6 gives examples of threads initiated from two separate sub-topics based on the discussion topic and instructions provided in Table 3 and 4.

Table 5. Example Sub-Topic: $CaCO_3$

(Student 1) Initial Sub-topic Post
Sub-topic: $CaCO_3$ *$CaCO_3$* *Cation = Calcium ion (Ca) $^{2+}$* *Anion = Carbonate ion $(CO_3)^{2-}$* *Type of bond:* *i. There is an ionic bond between (Ca) $^{2+}$ and $(CO_3)^{2-}$* *ii. The difference in Electronegativity between C and O is $3.5 - 2.5 = 1$. So the type of bond between C and O is a polar covalent bond.* *Name of the compound = Calcium Carbonate*
(Student 2) Response
Hi (Student 1), *Can you explain why the reason why your equation was $CaCO_3$ instead of $Ca_2(CO_3)_2$? The charges appear to cancel each other, is this the reason?* *Thanks* *(Student 2)*
(Student 1) Response
(Student 2), good question. Great guess too. *To be clear I am right because, when the charges are equal in quantity, they are not written. When charges are not equal, the charges are crisscrossed and written, for example:* *Ca^{+2} and Cl^{-1}* *Ca_1Cl_2* *$CaCl_2$*
(Student 2) Response
Thanks (Student 1). *In some cases it appeared that the charges due to them being equal cancelled each other but then I had an example where it reduces to its lowest terms. For example:* *Cation: (Sn) $^{+4}$ and Anion: (O) $^{-2}$* *Do the butterfly is* *Sn_2O_4* *SnO_2* *Am I understanding this correct?*
(Student 3) Response
Hey (Student 2), good job on the corrections but I think if you place the brackets down around the molecule it would make it easier to see what it is you need in the answer, But the first answer I feel is correct.
(Student 1) Response
(Student 2), that is right. It is certainly the correct ratio.

Continued on next page.

Table 5. (Continued). Example Sub-Topic: CaCO₃

(Student 4) Response
Hi (Student 1), *I believe I had asked you the same question as (Student 2) in class today during lab. If not close to it. I think what she is asking is if we are supposed to reduce to the lowest terms as if we were doing a math problem. For example 2/8 =1/4.*

Table 5 shows how the students help each other understand the concept of chemical formula and naming. In doing so, they are able to clarify certain details about chemical formulas, like when to put the subscripts and when to leave out or reduce them. Understanding these details about a chemical formula motivates the students to engage in the discussion. In the example given in Table 6, the students are learning at a faster pace and move on to the topic of molar mass and stoichiometry. These two examples exemplify the advantages of a clustered discussion board over a traditional discussion board.

Table 6. Example Sub-Topic: Mg₃(PO₄)₂

(Student 1) Initial sub-topic post
Cation= (Mg)²⁺ *Anion= (PO₄)³⁻* *Do the butterfly* *= Mg₃ (PO4)₂* *-Type of bond* *(Mg)²⁺ and (PO₄)³⁻ = Ionic bond* *P and O = the difference in electronegativity between P and O* *   3.5 -2.1 = 1.4 therefore the type of bond is polar covalent* *Name of the compound* *Magnesium Phosphate*

(Student 2) Response post
Hi (Student 1), *I am going to calculate the molecular weight of your compound Mg₃(PO₄)₂* *Mg:24.305*3=72.915g* *P:30.974*2=61.948g* *O:15.9994*8=127.9952g* *Molecular weight of Mg₃(PO4)₂=72.915g+61.948g+127.995g* *Molecular weight =262.858g* *How many grams are in 0.38 moles of Mg3(PO4)2* *Equation=262.858g/1mole*0.38 moles* *0.38 moles=99.88604 g*

Facilitation/Feedback

The role of the instructor in the clustered discussion framework is minimal and consistent with the "guide on the side" approach advocated by Mazzolini and Maddison (20, 21). Once the topic is supplemented with detailed instructions, the instructor is just there to make sure that the students are following the instructions, quality of the posts are maintained and incorrect information or misconceptions do not get propagated on the discussion board. Even if there is a post where a student has made a mistake, the instructor will stand back and allow time for other students to come in and help the student. The instructor will jump in only when the students are not able to help each other to arrive at the correct conclusion/answer.

Most students are usually nervous about being the first few to post on a discussion board. The best way to address this is through positive feedback on the forum and by repeatedly reminding the students that most of their classmates are in the "same boat as they are". It does help to respond to some of the good (correct) posts initially and let them know that is what a good post should look like. In my experience, after the first couple of weeks the students get over their initial inhibitions and start openly asking and accepting each others' help. This greatly improves the collaboration between students. On the opposite spectrum you may also have some over-enthusiastic students who post indiscriminately. To a great extent, this can be avoided by providing very clear instructions on the number and quality of posts. Often these students hold back on the indiscriminate postings after they see that they might lose points for not following the instructions and that they are not receiving any additional grades for doing so. This again emphasizes the importance of instructions and clear grading rubrics.

Conclusion

The traditional online discussion board starts with one initial post based on the discussion topic and then moves horizontally with the students adding new content and expressing their opinions, views, explanations and understanding of concepts. However, science is not always a subject that is amenable to long discussions based on views and opinions. This often makes it hard to conduct a traditional online discussion on a science topic that leads to actual learning. Furthermore, because of the structure and framework of the traditional discussion board, information gets lost in the hundreds of posts and threads that are not organized. Students often have to click through multiple long horizontal threads to follow the discussion. These factors discourage students from participating and many instructors refrain from including a traditional discussion board in the science curriculum.

A clustered online discussion board overcomes the above mentioned limitations of the traditional discussion board. Unlike the traditional discussion that moves predominantly horizontally, the clustered discussion moves both vertically and horizontally. The core of the clustered discussion board is the discussion topic that has a "multiplier effect". It forms the foundation on which students are able to create multiple initial sub-topic posts and hence the vertical movement of the discussion. These sub-topics posts then form the starting point for horizontal discussions. Thus in a clustered discussion board framework, the posts are automatically organized and categorized making it easier for the students to follow a discussion and find useful information. The detailed instructions that form an integral part of the clustered discussion board framework helps maintain the quality of posts and achieve specific learning goals. Students research, discuss, clarify, reason and explain the concepts in these discussions leading to concrete learning outcomes. They also develop their communication and collaboration skills by answering each other's questions, often in considerable depth. The role of the instructor in these discussions is limited and mainly follows the "guide on the side" model proposed by Mazzolini and Maddison (*20, 21*). In conclusion, the clustered online discussion board model helps overcome the limitations of the traditional discussion board while significantly reducing the load on busy students and instructors trying to keep track of long horizontal discussion threads.

References

1. Horn, M. B.; Staker, H. The Rise of K−12 Blended Learning. http://www.christenseninstitute.org/wp-content/uploads/2013/04/The-rise-of-K-12-blended-learning.pdf (accessed February 4, 2015).
2. Alammary, A.; Sheard, J.; Carbone, A. *Australas. J. Educ. Technol.* **2014**, *30*, 440–454.
3. Clark, I.; James, P. In *Proceedings of The Australian Conference on Science and Mathematics Education (formerly UniServe Science Conference)*; 2012; Vol. 11, pp 19–24.
4. Roberson, L. *J. Interpret.* **2015**, *24*, 1–16.
5. Reissmann, D. R.; Sierwald, I.; Berger, F.; Heydecke, G. *J. Dent. Educ.* **2015**, *79*, 157–165.
6. Rovai, A. P.; Jordan, H. *Int. Rev. Res. Open Distrib. Learn.* **2004**, *5*, 1–13.
7. Dixson, M. D. *J. Scholarsh. Teach. Learn.* **2012**, *10*, 1–13.
8. Martyn, M. *Educ. Q.* **2003**, *1*, 18–23.
9. Comer, D. K.; Clark, C. R.; Canelas, D. A. *Int. Rev. Res. Open Distrib. Learn.* **2014**, *15*, 26–82.
10. Premagowrie, S.; Vaani, R. K.; Ho, R. C. *Am. Int. J. Soc. Sci.* **2014**, *3*, 107–116.
11. Ka, H. *J. Nurs. Educ.* **2014**, *53*, 531–536.
12. Harman, K.; Koohang, A.; Harman, K.; Koohang, A. *Interdiscip. J. E-Learn. Learn. Objects* **2005**, *1*, 67–77.
13. Syh-Jong, J. *Educ. Res.* **2007**, *49*, 65–81, DOI:10.1080/00131880701200781.
14. Rivard, L. P.; Straw, S. B. *Sci. Educ.* **2000**, *84*, 566–593.
15. Paulisse, K. W.; Polik, W. F. *J. Chem. Educ.* **1999**, *76*, 704–707, DOI:10.1021/ed076p704.
16. Said, D.; Wanas, N. *Int. J. Comput. Sci. Inf. Technol.* **2011**, *3*, 1–14, DOI:10.5121/ijcsit.2011.3201.
17. Cobo, G.; García-Solórzano, D.; Morán, J. A.; Santamaría, E.; Monzo, C.; Melenchón, J. In *Proceedings of the 2nd International Conference on Learning Analytics and Knowledge*; ACM, 2012; pp 248–251.
18. Kay, R. H. *Br. J. Educ. Technol.* **2006**, *37*, 761–783, DOI:10.1111/j.1467-8535.2006.00560.x.
19. DeCristofaro, C.; Murphy, P. F.; Herron, T.; Klein, E. *Int. J. Arts Sci.* **2014**, *7*, 45–57.

20. Mazzolini, M.; Maddison, S. *Comput.   Educ.*   **2007**, *49*, 193–213, DOI:10.1016/j.compedu.2005.06.011.

21. Mazzolini, M., Maddison, S. In *Technology Supported Learning and Teaching−A Staff Perspective*; O'Donoghue, J., Ed.; Information Science Pub: Hershey, PA, 2006; pp 224–241.

Chapter 6

An Online Research Methods Course at Georgia Southern University

C. Michele Davis McGibony*

Department of Chemistry, Georgia Southern University,
Statesboro, Georgia 30460-8064, United States
*E-mail: mdavis@GeorgiaSouthern.edu

The Research Methods course at Georgia Southern University was first established in 1998 as a supporting pre-requisite to our upper level courses. The two main topics covered were chemical literature and data analysis. The evolution of the internet, significant changes in the availability of computer-based tools for data analysis and literature searching, and increased expectations of students in upper level courses necessitated an update to this course, which was revised in academic year of 2008–2009. The result was a course for sophomore and junior chemistry majors with material focusing on computer skills, laboratory skills, chemical literature research skills, and scientific ethics, as well as careers and internships in the chemical field. This course was first implemented in the spring term of 2010 as a hybrid course (mixture of face-to-face and online) which included evaluation surveys at the end of each unit. Based on the feedback from students and the instructor, the course has since been offered as a wholly online course. This book chapter will describe the evolution, the content, and the pitfalls of an online research methods course.

Georgia Southern University is the state's largest institution of higher education south of Atlanta. Our university has over 120 degree programs at the baccalaureate, master's, and doctoral level. We are a classical residential campus with more than 20,000 students from 48 states and 89 nations. For over 100 years, the university fosters a culture of engagement that bridges theory with practice, extends the learning environement beyond the classroom, and promotes student growth and future success. Our university tagline is "large-scale small-feel" and Georgia Southern University is well-known for its outstanding professors and friendly personal environment for all students. Our institution is committed to many high impact practices to engage students and create life-long scholars such as the First Year Experience program, learning communities, service learning opportunities, and a required course, Global Citizens.

The Department of Chemistry is housed within the College of Science and Mathematics. This department offers a modern chemistry program leading to BS certified degrees by the American Chemical Society. Our department is ranked in the top 25 producers of ACS-certified BS chemistry majors (1) nation-wide and our Student Chapter has been recognized by the ACS for the past twelve years (2). The faculty is committed to providing a student-centered environment to develop each student as life-long learners and members of the scientific profession. This includes a well-balanced curriculum that consists of strong and innovative instruction accompanied by modern laboratory methods, technologies, and collaborative projects.

Most chemistry programs (3–7) across the nation require their majors to obtain skills in chemical literature searching, and this skillset is required for degree certification by the American Chemical Society (8). Georgia Southern University is no different in their expectations, and all our BS and BA degrees are ACS-certified. During the 1980s and early 1990s, students earning a BS degree were required to earn two credit hours in "Chemical Literature". This course focused on techniques and tools for effectively searching all forms of chemical literature including Beilstein, chemical abstracts, CRC Handbook of Chemistry and Physics as well as library resources. Another requirement during this time was a "Computing in the Sciences" course taught in the Department of Mathematics and Computer Science. This course focused on spreadsheets, word documents, databases, and other technologies available.

For the academic year of 1994–1995, the entire university underwent conversion from quarters to semesters. This entailed an entire curriculum reform. At this time, the chemistry department at Georgia Southern University decided this was the perfect time to combine the literature and computing courses and add additional material valuable to all students majoring in chemistry. The new course was given the title "Research Methods". Learning outcomes for this course included computer literacy (Word, Excel, and Powerpoint), data analysis, chemical literature, and visualization of molecules. This course was taught with a novel approach of "work at your own pace" with instructor mini-lectures and one-on-one assistance from the professor during class times. The material was prepared for students in their sophomore year. In addition to this change in the curriculum, the chemistry department at Georgia Southern University also integrated two Junior and Senior seminar courses that were one credit hour each.

Table 1 contains a summary of goals and majors topics for these three courses taught prior to 1995. The Junior seminar course focused on all the available careers for chemistry majors and how to write an effective resume, cover letter, and personal statement. Senior seminar focused on a capstone research project and allowed students to gain presentations skills.

Table 1. Goals and Major Topics for All Three Courses Taught at Georgia Southern University Prior to the Consolidation and Creation of Principles of Chemical Research Methods

CHEM 2031: Research Methods in Chemistry	Goal: To use various computer programs effectively to act as a scientific professional. Topics include: data analysis, statistical analysis of data, internet and library searching of chemical literature, and 3D visualization of molecules.
CHEM 3610: Junior Seminar	Goal: To introduce students to the skills necessary to be a competitive chemical professional and possible career paths. Topics include summer research programs, resumes, cover letters, professional schools, graduate schools, industry. Guest speakers introduce various topics.
CHEM 4611: Senior Seminar	Goal: To enhance the skills and abilities to function as a scientific professional. Topics include understanding the chemical literature, discussion of chemical literature, and presentation of chemical literature in a professional manner.

This combination of classes, Research Methods, Junior Seminar, and Senior Seminar, worked very well for several years. However, as time passed and technology became more integrated into the daily lives of students, students and faculty became dissatisfied with these courses. From the student perspective, they viewed themselves as computer and technology savvy. They could not see the real life application of these skills but only saw busy work for no reason. From the faculty prespective, the main problem was the skills from the research methods course were not transferable to other courses. For example, in the old research methods course an assignment using MathCad was introduced because MathCad was a program used extensively in physical chemistry, but students didn't use the program again until two or three semesters later. Another problem that arose from the seminar classes, from the faculty perspective, was that these classes were not taught as part of a faculty member's teaching load but as extra work for each professor in the department.

From this issue came a new idea, to create one course and call it "Principles of Chemical Research". This course would be aimed at sophomore to junior level students with materials from all three courses (research methods and both seminars) for a combination of 3 credit hours that would be taught in load for the faculty member. The course would be taught in a module format, but assignments would have concrete, real-life applications. The materials taught in this course would be the skills students needed right now; more advanced applications for upper level students would wait until they needed the skills. The MathCad assignment described earlier and others like it were removed from the new course. With the module format, this course could easily be a hybrid or completely online course.

Table 2. Listing of Modules for Newly Designed CHEM 2030, Principles of Chemical Research Methods

Module 1: Introduction to CHEM 2030 an Online Course
Module 2: Laboratory Safety and Equipment
Module 3: Scientific Ethics
Module 4: Chemical Literature
Module 5: Scientific Writing
Module 6: Microsoft Word and Chemical Structures
Module 7: Molecular Modeling and Visualization
Module 8: Microsoft Excel and Data Analysis
Module 9: Careers for Chemists
Module 10: Summer Research Opportunies

The new course, Principles of Chemical Research, was taught in the spring term of 2010. All modules included in this course can be found in Table 2. Many of the modules were a compilation of materials taught in the previous three courses, but the modules over laboratory safety and ethics were added to this course; the module on careers was changed significantly. This was a hybrid course with approximately 70% online delivery, and almost all students enrolled were in the spring semester of their sophomore year. After the completion of each module, a survey was given that asked what were the "best" and "worst" parts of the modules and what would they like to see change for the future. According to these evaluations, most students enjoyed this course, and a few even stated it was their favorite science course yet. The students did offer insightful comments about what was needed in order to teach the material in a fully online format. Two main issues were health professional careers were lacking and that many assignments seemed too long and needed to be in smaller pieces. Students also commented that they did not see the purpose of including the laboratory equipment module. Using this information, CHEM 2030 went completely online in the summer term of 2010 with the addition of health related careers in the careers module and the addition of a research opportunities module. The laboratory equipment and safety module was also revamped in order to demonstrate to students this knowledge, at this time in their career, was necessary and relevant. The content of this course has evolved over time, but the ten modules in Table 2 are still currently used for this course, and the basic layout of each module is described herein.

The first module is an introduction to the course which includes the syllabus, the course schedule with due dates, and tips on taking an online course since this is often the first course of this type in a student's college career. This module contains a quiz and two assignments. The first task is to post in the discussion board a basic introduction of themselves and a current photo in order to create a sense of online community. The second assignment is to use the email within the course platform (Folio at Georgia Southern University) to send the instructor a message. The entire purpose of this module is to aquaint the student with the online environment and to communicate key course policies regarding grading and late assignments.

The laboratory safety and equipment module focuses on basic hazard communications information like MSDS (Material Safety Data Sheets), proper chemical labeling, and waste handling as well as the use and care of commonly encountered laboratory equipment such as micropipets, volumetric flasks, and analytical balances. This module delivers the information via voice-over powerpoints, reading, and interactive videos. For assessment, two quizzes and one assignment are required. The quizzes are timed, but each student has two attempts for each quiz in order to maximize the retention of material and the student's grade.

The scientific ethics module discusses plagiarism in its various forms, the need for proper and correct citation of information, research misconduct, and many other topics on scientific ethics. Students are given the document "On Being a Scientist: A Guide to Responsible Conduct in Research" (9) for background information on this topic. This module also uses case studies to discuss ethical problems and the philosophy of scientific inquiry. These case studies come from "The Ethical Chemist: Professionalism and Ethics in Science" (10) as well as current news articles about science misconduct in the USA and around the world. This module contains two quizzes which are timed with two attempts, one focusing on the article "On Being a Scientist" and the other on plagiarism and ethical behavior in general as well as rules specific to our institution.

The third module in the sequence focuses on scientific writing, publication, and chemical information. The ACS Style guide was utilized for students to learn how to write lab reports, scientific papers, and scientific presentations including correct citation methods. The differences between primary, secondary, and tertiary literature is explored, and most importantly online searching via our institution library and the internet. This module contains two quizzes which are timed with two attempts and one assignment which was adapted and modernized from the former research methods course.

The Microsoft Word and Chemical Structures module teaches students to produce professional quality research reports or summaries. The main goals of this module are for students to create chemical structure in ChemSketch, create tables in Microsoft Word, and insert these as well as other objects into a document with appropriate figure and table captions. This module contains one assignment.

Molecular visualization allows the students to view, rotate, modify, and analyze chemical molecules in three dimensions using a specific program of the instructor's choice. Programs used in recent years have been WebMO (11), RasMol (12), JMol (13), and Protein Explorer (14). All these programs are free and available for instant download. The only assignment for this module allows the students to view and analyze various molecules with a range of sizes within the chosen program. In the past offering of this course, caffeine, cholesterol, hemoglobin, silk, and HIV protease were used.

The next module in the course is one of the most important and most difficult for the students. This module focuses on Microsoft Excel, which is the most prevalent software for data handling and analysis. The goal of this module is for students to create a graph in Excel with a trendline and linear regression table as well as analyze data statistically and use the information to solve a problem. Intially this module contained one large assignment, but after a few offerings it was obvious that the students needed smaller "bite-sized" portions of this material. The course was recently revised to break up two large assignments. There are currently four assignments in this module, each focusing on a different task: Statistics, Graphing, Linear Regression and Residuals, and Data Analysis. The final data analysis assignment allows students to use the skills they learned in the previous assignments to solve a "real" problem. The problem and data given vary from year to year depending on instructor's preference. Most recent problems involve determining the amounts of calcium in breakfast cereal, the caffeine in various beverages, and cholesterol in patient blood samples. Every assignment allows the students to enter data in a spreadsheet, calculate the mean with standard deviation, create a graph of standards, add a trendline, and make the chart suitable for a professional publication. Finally, the chart is used to calculate an unknown quantity using the standard curve created and the regression information.

The next module explores various career options for students majoring in chemistry provides the tools needed to qualify for the positions of specific interest to individual students. These modules introduce students to post-baccalaureate studies such as graduate school within the field, medical options (medical, dental, physician assistant, podiatry, anesthesiology assistant, physical therapy, etc), high school teaching, governmental positions, and many others and the how and when to apply for these positions. It is always interesting each year for the instructor to read the students' reactions when they discover that graduate schools within the field have free tuition and a stipend. This is perhaps the most eye-opening module for most students. This module contains two quizzes (timed and two attempts) and two assignments. The assignments are referred to as "Plan A" and "Plan B". Most students arrive at college with one single career plan. Often their plans evolve over time or the initial plan is not possible with their given skillset. This assignment allows students to research their ideal career and what it will take to get them where they want to be; it also forces them to explore many alternative careers. Figure 1 below contains the online assignment for this module.

As a continuation of this module, information about interviewing and the documents necessary to obtain a specific job, internship, fellowship, or other position is delivered. Students watch videos and presentations about interviewing and the differences between a resume and a curriculum vitae. This module contains one quiz and three assignments (draft resume, cover letter or personal statement depending on Plan A, and a final resume). Students receive feedback on their resumes from the instructor and our institutional career service center. At the end of this module, each student has a well crafted resume that is posted on our EagleCareerNet website for potential employers.

The final module introduces students to the many different research opportunities available within our department and college as well as outside the university. There is a plethora of data that indicates the earlier students engage in undergraduate research the more likely they are to continue to excel in their program of study (15). It is imperative to provide students with knowledge and access to the many types of research opportunities available. Students become familiar with NSF REU programs, the McNAir program, various internships, institutional research credit, and how and when to apply for these positions. Students review a voice-over powerpoint presentation and take an online quiz. Normally after the completion of this particular module the instructor and the director of undergraduate research in our college receive many emails from students interested in obtaining hands-on research experience.

Most students have an idea of what they want to do after graduation, and I hope after viewing all presentations and many of the weblinks provided everyone has an idea about what type of career they want to have in the future. Your "Plan A" is your primary plan after graduation if all goes well in terms of GPA and experience. Please follow the instructions to complete this assignment for your primary plan.

For the Plan A assignment, you must answer these questions in complete sentences and proper grammar.
1. Describe your career aspirations once you leave GSU.
2. Describe how earning a chemistry degree will help you achieve this goal.
3. Describe your expectations of what a typical day in your life, including how many hours/day you will be working.
4. What will your salary be?
5. What education is necessary after you leave GSU? How long will it take?
6. Outline in table format the classes you need to take to graduate on time. Highlight the other things you need to do over the next 1-2 years to achieve your goal (take entrance exams, shadow, intern, etc.) A sample table is attached.

Figure 1. Module 7: Plan A Assignment.

As with any course and any delivery format there are advantages and disadvantages, and Principles of Chemical Research is no exception. From a student perspective, most students enjoy this course and feel that they gain valuable knowledge and skills that they will use in the future. Many students enjoy the online format and the flexibility to work ahead and around other course exams and deadlines. This usually appeals to the highly motivated student. The favorite modules over the last five years are Careers for Chemists and Summer Research Opportunities; the least favorite are the Microsoft Excel and Laboratory Equipment and Safety. Students are often frustrated with automatic grading in our platform (Folio). Most short answer and all long answer questions have to be manually graded by the faculty member, so the intial grade students see for a quiz is much lower than their actual grade. It often takes many announcements via the news board and discussion board to clear up this issue. In the past two offerings of this course, new questions about this process have been included in the introductory module. From a faculty perspective, the flexibility is wonderful, and since this course has been taught numerous times by four different faculty members there is a wide range of questions and module assignments to choose from when creating a specific course. The two main issues faculty face while teaching this course are communication with students and expectations of students. Without regular face-to-face interaction, the nonverbal gestures and facial expression that constitute communication are removed. This can often lead to misunderstandings, context errors, and frustration between students and between students and the professor. Another issue is communication timing—therefore, each instructor must set up guidelines for communication during the course introduction. At Georgia Southern University, our faculty state in the course syllabus that we will check the folio course at least twice a day during the regular work week and at least once on the weekend. The reality is we check in on the course much more than stated to surpass student expectations. When teaching this course, instructors must realize that self-discipline of the individual student is required to do well in an online course. Since these students do not have regular face-to-face interactions with the professor or other students who can help with reminders and accountability, the student must be self-reliant and vigiliant about due dates and times. In order to motivate students in the couse, the instructor usually sends a message to all students at least three times a week reminding them of deadlines, encouraging them to submit their work, and posting helpful hints if students have already asked questions about a particular assignment.

Looking toward the future, the course will continue to evolve and change. Currently, a new module is in the works focusing on the 12 Principles of Green Chemistry in conjunction with the American Chemical Society and the Office of Sustainability at Georgia Southern University. Additionally, faculty who teach this course are working with our Center for Online Learning to create applications within each module that are more mobile-friendly for students who use tablets and other mobile devices (*16*).

As the first completely online course offered in the Department of Chemistry at Georgia Southern University, Principles of Chemical Research has been quite successful. This course gives students early in their college career plenty of information about career options and undergraduate research as well as teaching them skills in scientific writing, scientific ethics, basic laboratory safety, basic laboratory equipment, molecular visualization, and data analysis. These skills are necessary in many junior and senior level courses and within their own individual research projects. Using the information presented in this book chapter, this type of course could be implemented successfully at any university across the globe.

Acknowledgments

This course has been a collaborative effort among many faculty members in the chemistry department at Georgia Southern University. Brian Koehler and Jim LoBue were the main authors of the course materials for the original research methods course, and Shannon Davis was the first instructor of the new hybrid course. Other instructors who have contributed to the question data bank and additional assignments are Allison Amonette and Mike Hurst. Feedback from the over 450 undergraduate students at Georgia Southern University has been invaluable in improving this course.

References

1. Annual Reports of Earned Bachelor's Degrees in Chemistry. American Chemical Society, Committee on Professional Training. http://www.acs.org/content/acs/en/about/governance/committees/training/reports/degreesreport.html (accessed February 26, 2015).
2. Student Chapter Award Recipients. American Chemical Society. http://www.acs.org/content/acs/en/funding-and-awards/awards/community/sachapter.html (accessed February 26, 2015).
3. Matthews, L. *J. Chem. Educ.* **1997**, *74*, 1011–1014.
4. Abrash, H. *J. Chem. Educ.* **1992**, *69*, 143–146.
5. CH 350 Chemical Literature Syllabus. Western Oregon University. http://www.wou.edu/las/physci/ch350/ch350syl.htm (accessed February 26, 2015).
6. CHEM 184/284 Syllabus. University of California, Santa Barbara . http://guides.library.ucsb.edu/chem184 (accessed February 26, 2015).
7. CHM 513 Chemical Literature Syllabus. Purdue University. https://www.chem.purdue.edu/academic_programs/resource-room/docs/syllabi/CHM%20513_Syllabus_Sp2013.pdf (accessed February 26, 2015).
8. *Undergraduate Professional Chemical Education*; American Chemical Society, Committee on Professional Training: Washington, DC, 2008, 14.
9. *On Being a Scientist: A Guide to Responsible Conduct in Research*, 3rd ed.; The National Academies Press: Washington, DC, 2009.
10. Kovac, J. *The Ethical Chemist: Professionalism and Ethics in Science*; Prentice Hall, 2004.
11. WebMO. http://www.webmo.net/ (accessed February 26, 2015).
12. RasMol and OpenRasMol. http://rasmol.org/ (accessed February 26, 2015).
13. Jmol: an open-source Java viewer for chemical structures in 3D. http://jmol.sourceforge.net/ (accessed February 26, 2015).

14. JMol Protein Explorer 0.5. http://chemapps.stolaf.edu/pe/protexpl/htm/ (accessed February 26, 2015).

15. Kuh, G. D. *High Impact Educational Practices: What They Are, Who has access to Them and Why They Matter*; Association of American College and Universities, September 2008; pp 13–23.

16. McGibony, C. M. D. *Research Methods at Georgia Southern University: Where Are We Now?*; Abstract # 420; SERMACS: Nashville, TN, October 17, 2014.

Chapter 7

PSI4Education: Computational Chemistry Labs Using Free Software

Ryan C. Fortenberry,[1,*] Ashley Ringer McDonald,[2,*]
Tricia D. Shepherd,[3] Matthew Kennedy,[4] and C. David Sherrill[4]

[1]Department of Chemistry, Georgia Southern University,
Statesboro, Georgia 30460-8064, United States
[2]Department of Chemistry and Biochemistry,
California Polytechnic State University,
San Luis Obispo, California 93407, United States
[3]Department of Chemistry, St. Edward's University,
Austin, Texas 78704, United States
[4]Center for Computational Molecular Science and Technology,
School of Chemistry and Biochemistry, and
School of Computational Science and Engineering,
Georgia Institute of Technology,
Atlanta, Georgia, 30332-0400, United States
*E-mail: rfortenberry@georgiasouthern.edu (R.C.F.);
armcdona@calpoly.edu (A.R.M.)

Computational chemistry is a significant tool in contemporary research, but little of it is employed or taught as part of the standard undergraduate chemistry curriculum. This has largely been due to the cost of hardware and software, a lack of computationally specific technical expertise on the part of the average educator, and little freely-available computational chemistry curricular material. PSI4Education is a team of individuals hoping to change this. In this chapter, we discuss how the WebMO graphical-user interface and PSI4 suite of quantum chemical programs, which are both available at no cost, can be installed and utilized for educational purposes. More importantly, however, PSI4Education has developed a set of ever-growing, vetted laboratory exercises to be employed in the chemistry classroom and laboratory. These are available free of charge for download from our website: http://www.psicode.org/labs.php. The labs are created in a flexible manner to be used as-is or tailored for particular use by the educator, and they come with answer keys and various other helpful instructional tools. We welcome further contribution from the community and believe that the next generation of chemists must have at least basic competency in computational chemistry as they do in synthesis, instrumentation, and all of the other traditional aspects of chemical education.

Introduction

Beginning with the initial undergraduate-level courses taught at most colleges and universities, mathematical models are necessary to describe chemical phenomena. These are often relatively simple algebraic relational models, but the Arrhenius equation and Charles's Law, as examples, are essential for providing students with necessary levels of chemical under- standing. The advent of quantum mechanics greatly improved chemists' understanding of the underlying physics dictating the observed chemistry, but true quantum chemical modeling is nearly impossible to perform with pencil and paper for any systems of real interest (1–3), largely placing its usage out of the hands of most chemistry students. Luckily, the introduction of computers into chemical research has greatly improved the depth of employed models, rendering computational chemistry a necessity for modern research. The flexibility that molecular modeling provides in detailed analyses of orbital occupations, reaction dynamics, and potential energy scans makes its inclusion in the modern set of research tools a natural extension. This flexibility can also make computational chemistry a fantastic educational tool (4), especially for more complicated concepts such as visualizing molecular orbitals, symmetry operations, and electrostatic potential mapping, but very few undergraduate students are routinely exposed to computational chemistry before graduation. Hence, contemporary chemical education possesses few if any concepts of computational chemistry even though this tool is opening new areas of chemical research (5, 6). Students must learn at least basic computational chemistry skills in order to become the productive chemists of tomorrow.

Forays into the use of computers and computational chemistry in the undergraduate chemistry curriculum have been reported in the literature since the 1990s (7–9) when computer literacy was on the rise and the cost of computer hardware was declining. As the subfield of physical chemistry was increasingly populated by trained computational chemists, computational chemistry began to emerge more frequently in the undergraduate chemistry classroom and laboratory (8, 10, 11). In fact, a large portion of an issue of the *Journal of Chemical Education* (Vol. 73, Iss. 2) was dedicated to the use of computational chemistry in the undergraduate classroom in 1996, and the first three volumes (2005–2007) of *Annual Reports in Computational Chemistry* (a yearly publication of the American Chemical Society) devoted an entire section to topics in chemical education. Around the same time, research dollars were being granted to provide the necessary hardware and manpower to examine how computational chemistry could be developed for undergraduate education and even research (*11*). Additionally, pioneering computational chemists found new ways to communicate the minimum of necessary technical Linux/Unix competencies to students while producing positive learning outcomes for the physical chemistry classroom (*12*). As such, the use of computers and computational chemistry as an educational tool has ballooned in recent years (*13–17*).

In spite of an increase in computer usage for research and education within chemistry, the amount of computational chemistry education provided in the standard undergraduate curriculum is still almost nonexistent, especially at small and underserved colleges and universities. The reasons (*4*) for this are threefold: 1) technical expertise, 2) financial expenditure, and 3) vetted computational educational activities. A lack of deep-knowledge computer literacy by both the faculty and the students has long hampered further development of computational chemistry as a teaching tool (*12*). Most computational chemistry programs have been written by computational chemists with a bent towards the necessary computer science aspects of their careers. Hence, many packages require expert-knowledge for installation and even usage. Many faculty who are traditionally-trained chemists and not computational or theoretical chemists are not confident enough to engage in this type of instruction. Additionally, most students have no previous knowledge or experience with such high-level programming, forcing the instructor to spend large amounts of time teaching the students how to use the program and not learning the desired chemical concepts.

On the other hand, many companies have produced relatively easy-to-use and easy-to-install computational chemistry programs, but these can cost in the range of thousands of dollars per license per year, well above what most departments can afford for their numbers of students. Such software is often viewed as a risky purchase anyway since most faculty are not familiar with the education potential for these programs. Beyond the cost of the software, hardware concerns are also of financial consideration. Basic and workable computer hardware can be obtained readily today, as most colleges and universities purchase large numbers of computers and have information technology (IT) personnel on staff. However, if more advanced techniques are to be employed and exposed to the students, access to quality hardware can still be an issue. Remote supercomputing and cloud-based resources are beginning to show promise as a means of alleviating such concerns (*18*), but this represents a very new path for computational chemistry education.

Even if hardware is available, software is purchased, and the educator is competent and capable of setting up the quantum chemical programs, the exercises and assignments using these resources must still be developed. This can take several hours to several weeks to create, evaluate, and implement, representing a significant time-investment by the educator, especially if he or she is building these ideas and concepts from scratch in isolation. Luckily, the world is beginning to change on all of these fronts.

PSI4Education

Some modern, state-of-the-art software is actually available for download free of charge, immediately reducing the operating costs of computational chemistry as an education tool. The WebMO (*19*) graphical-user-interface (GUI) is a free, easy-to-use, HTML- and Java-based program that can be attached to various computational chemistry engines. Higher functionality of this program does require paid license upgrades on the order of a one-time purchase of $1,000 (US) to $2,000 (US; depending on version), but these are relatively inexpensive compared to other computational chemistry GUI licenses which can, again, cost upwards of $2,000 (US) per student per year. Additionally, the PSI4 (*20*) suite of computational chemistry programs is a free and open-source package that is capable of high-level quantum chemical computations and has been utilized extensively for high-level computational chemistry research (some of it even performed by undergraduate students) (*21–26*). It is developed by both domestic and international partners at universities including Virginia Tech, Georgia Tech, the University of Georgia, Ataturk University, and Emory University among others. The PSI code has existed in various forms since the mid-1970s with PSI4 development beginning in 2009. Its development has largely been funded by the National Science Foundation and among some communities is becoming one of the most-used quantum chemistry programs available. Hence, the marriage of WebMO and PSI4 promises low-cost, or even no-cost, tools that can be used as part of undergraduate education.

There is some initial setup of the WebMO and PSI4 software required on the part of the instructor. The PSI4 program currently only runs on Linux/Unix or Mac OS X operating systems. However, for Linux machines a binary distribution is now available, as described at http://www.psicode.org/psi4manual/master/conda.html. The binary distribution includes all required dependencies and is updated nightly; the PSI4 installation can be updated automatically with a single command. WebMO can be installed on Windows, Linux/Unix, or Mac OS X operating systems. However, the free version of WebMO requires installation on the same machine that is used to run the computations, and hence to be used with PSI4. WebMO must be installed on a Linux/Unix or Mac system (unless the WebMO Enterprise version is purchased). Installation of WebMO on a Linux/Unix machine is fairly straightforward for anyone familiar with the basic features of Unix, and is assisted by detailed instructions on the WebMO website and a provided setup script. WebMO must be installed on a machine that provides a Web server, since students interact with WebMO via web browsers. Configuration of the Web server to accommodate WebMO is also described in the WebMO setup directions. Although some initial setup in UNIX is required by the instructor or the local IT staff, students interact with WebMO exclusively through a browser-based GUI that is very easy to learn and to use.

The authors are working within the PSI4 community to develop laboratory exercises and educational tools that make use of the WebMO program and its GUI linked to PSI4. Our effort is called PSI4Education, and our aim is to create meaningful laboratory exercises to increase students' exposure to computational chemistry and to provide the technical expertise to aid in the installation and setup of the free WebMO/PSI4 software.

While there are several other free quantum chemistry packages that currently link with WebMO, including GAMESS, MOPAC, and NWCHEM (as well as TINKER and Quantum Espresso for dynamics), we are limiting our free and open-source lab manual to the free and open-source PSI4 program. PSI4 is unique in its exceptionally efficient density-fitted codes which significantly speed up any computations, especially for geometry optimization, which is important in all of the lab activites. Additionally, PSI4 can perform symmetry-adapted perturbation theory (SAPT) computations on molecular fragments, open-shell electronically excited states with coupled cluster theory, and coupled cluster multi-reference computations, to name a few unique features. These diversify the potential for PSI4Education exercises to be used for a variety of different types of future lab activities. Finally, PSI4 is, again, open-source, allowing expert users to modify the code in order to include new features if desired for pedagogical reasons. In fact, the top level or "driver" portion of PSI4 is written entirely in Python, and basic extensions are therefore easy to implement because Python is easy to learn and allows rather complex tasks to be accomplished using only a few lines of code.

Several books exist which present computational and quantum chemistry exercises for student use (27–29). These books vary in their focus, complexity, and level of instruction. For example, *Exploring Chemistry with Electronic Structure Methods* by Foresman and Frisch (27) gives detailed instructions and exercises for exploration of the commercial Gaussian® family of programs (30). The scope of PSI4Education differs from these other materials in its accessibility, flexibility, pedagogy, and expense. This project is open-source and welcomes continual contributations and improvements. Course-ready materials are provided with instantaneous access online and can be immediately implemented in the classroom. While these activities are designed to be instructional, they are written in such a way as to not only enhance computational chemistry skills, but also invite students into the research process.

PSI4Education released our first open source lab manual in August 2014, and it is freely available for download at www.psicode.org/labs.php. It is our hope to continue to add to this repository, and it will always be free of charge just like PSI4. The PSI4Education labs are protected by a Creative Commons Attribution-NonCommercial 4.0 International License, but as long as for-profit usage of the materials is not being promoted without consent from the team members, the labs are available for any educational purpose.

Every lab exercise published is titled in the form of a research style question in order to engage the student in the research process and hone their problem solving skills. Each exercise is complete with student instructions, a worksheet, and an answer guide for the educator which includes a set of learning objectives, an estimate of the expected time to complete the procedure, and a summary of the lab in addition to the proper results for the student exercises. These labs have all been vetted and tested by various members of the team and even implemented in many of our own courses. Potential pitfalls and issues are also mentioned and discussed in the answer guide such that the instructor is aware of areas where students may struggle or have issues as the learning process takes place. The labs are published in the Google Documents format of Google Drive. As such, the educator retains a tremendous amount of flexibility in what he or she may choose for the student to do. For instance, the educator may download the exercise to a word processor and remove the worksheet portion so that the students may write a lab report. The educator may choose to use only a portion of the exercise, or use an entire lab as-is. In any case, the educator can tailor the exercise for the students in his or her classroom. Additionally, these labs can be linked to typical online course administrative software such as Moodle, Desire2Learn, Blackboard, and similar programs. While the labs are freely accessible, the answer guides have restricted access in order to minimize the chances of cheating or other improper conduct from the students. However, an e-mail from the educator to the PSI4Education team from his or her university email account is enough for access to be granted.

Currently, our labs come in two categories: basic and advanced. Since most of the exercises developed thus far have been produced for the typical second semester physical chemistry course, i.e. quantum chemistry, most of the labs created have a bias for this course. Currently, labs related to molecular orbitals, symmetry, non-covalent interactions, theory/basis set considerations, and rotational spectroscopy have been produced. However, the advanced labs can be included in any chemistry course. The use of molecular orbitals may be of particular interest to organic chemistry, as an example. Future labs will include activities from across the chemistry curriculum. The basic labs are currently designed for introductory chemistry concepts that are often discussed in first-year undergraduate courses but can be applied to other discussions in more advanced settings where applicable. As of now, these include an initial example exploring the radius of an atom, giving students a visual description of how chemists classify the size of submolecular species.

Additionally, there exists other documentation for "Getting Started". This includes a tutorial designed for all new WebMO/PSI4 users regardless of education level. It shows students basic but necessary concepts including how to build molecules within the GUI and how to run computations. A discussion of customizing labs (as mentioned previously) is also given within the Getting Started section as well as a detailed set of instructions in how to setup and install WebMO and PSI4. Even though PSI4 requires installation on a Unix-based operating system such as Linux or Mac OS, the detailed instructions should be straightforward enough for most educators to install and link the programs. Additionally, the PSI4Education team is always available for assistance since we have all installed and maintain our own WedMO/PSI4 setups at our respective universities.

What It Looks Like

Since the students (and the instructor, besides installation) will only see the WebMO GUI, the WebMO website (http://www.webmo.net) provides a fantastic description of the visuals for their program including screen shots, troubleshooting, and tutorials. However, some issues are intrinsic to the use of WebMO with PSI4. To address these and provide an example as to how the exercises proceed, the lab entitled "Is C_3H^+ present in the Horsehead nebula?" is broken down and discussed below. In this lab, the students explore the astronomical molecular rotational lines observed in the Horsehead nebula photodissociation region and what the carrier of these features may actually be (*31–34*).

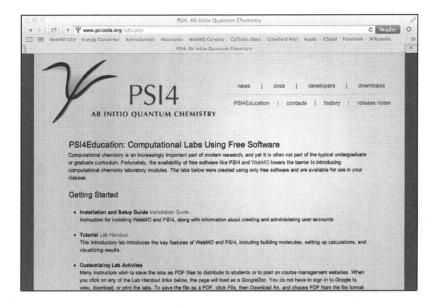

Figure 1. The PSI4Education website. (Printed with permission from www.psicode.org.)

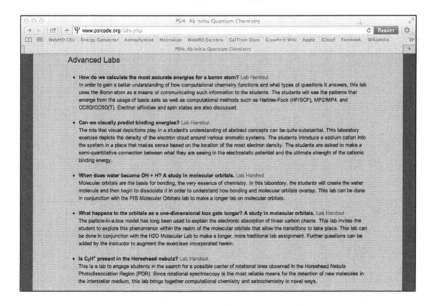

Figure 2. The advanced labs section on the PSI4Education website. (Printed with permission from www.psicode.org.)

On the PSI4Education website (Figure 1), the user can scroll down to the "Advanced Labs" section (Figure 2) and select the "Lab Handout" link at the bottom of Figure 2.

Is C_3H^+ Present in the Horsehead Nebula?

Developed by Ryan C. Fortenberry[1]

Rotational spectroscopy is the most conclusive way in which molecules can be detected in the interstellar medium, circumstellar envelopes, and, even, the atmospheres of various extrasolar planets. The reason for this comes from the clear progressions of transitions resulting from the

$$\Delta J = \pm 1 \tag{1}$$

selection rules. Additionally, the pure rotational spectra of most molecules are fairly straight-forward to understand. From the second-order fitting of the rotational energy level function,

$$E_J = 2B(J+1), \tag{2}$$

given in Pety and coworkers,[1] the primary rotational constant, B, for a linear molecule can be straightforwardly derived with only one energy level. Remember that the rotational constant is a geometric parameter inversely proportional to the moment of inertia, I. This value is simply defined as:

$$I = \Sigma\, m_i x_i^2. \tag{3}$$

However, the value of E_J derived as such from Eq. 2 is not as accurate as it could be. Since the molecule is spinning, the energy is affected by centrifugal distortions. To account for this, a quadratic term can be derived to give an equation a more correct form given as:

Figure 3. The first page of the Horsehead nebula laboratory exercise.

From the PSI4Education webpage, the user is linked directly to the Google Document for this lab as shown in Figure 3.

Computational Rotational Spectra

Name _____ Date _____

Lab Partner _____

Lab Partner _____

Part A

C_3H^+ B-type rotational constants (in GHz)

HF/STO-3G:_____

HF/6-31G(d):_____

HF/cc-pVTZ:_____

HF/cc-pVDZ:_____

B3LYP/cc-pVDZ:_____

MP2/cc-pVDZ:_____

Figure 4. The worksheet portion of the Horsehead nebula laboratory exercise.

The laboratory exercise is then laid out with an introduction and a procedure. The optional worksheet begins on the first page immediately following the procedure. This is depicted in Figure 4.

Figure 5. The job manager window for WebMO. (Printed with permission from reference (19)).

The WebMO GUI interface is easy-to-use and fairly intuitive. The instructions in the lab procedure are also step-by-step in order to provide the easiest means of execution for the student. Again, the instructor may choose to modify these per his or her desires. After logging in to WebMO, the student will come to the "Job Manager" window (Figure 5). By clicking "New Job", in the upper-left-hand corner, the student is taken to a screen where he or she can input the desired molecule.

This input is most readily done by building the molecule within the viewer to create the structure shown in Figure 6. Again, the PSI4Education tutorial in the "Getting Started" section discusses how to do this, as does the WebMO website. The molecules may be created as Lewis structures with mulitple bonds, but students often struggle with the creation of radicals and ions since the standard Lewis structure rules are no longer complete. The "bonds" given in the WebMO GUI are simply a measure of distance. Quantum chemically, bonds are just probability densities and actually do not mean anything to WebMO or the PSI4 engine. Only the position of the atoms and the number of electrons are important. Hence, a molecule drawn in WebMO with a proper Lewis structure or simply with points for the atoms will perform the same way as long as the number of electrons are equal. Regardless, the desired coordinates for the atoms and a best first guess as to the molecular structure is created in the "Build Molecule" window.

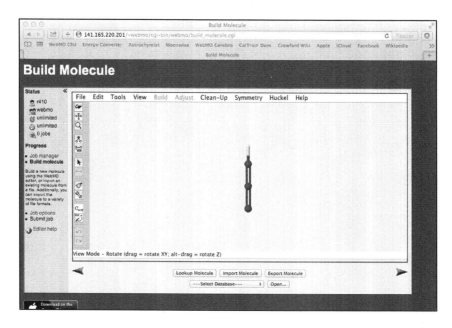

Figure 6. The "Build Molecule" window. (Printed with permission from reference (19).)

By clicking on the right arrow at the lower-right-hand corner, the built molecule can then be prepared for computation in the "Configure Job" window. The proper selections are given in the laboratory exercise, but the final product will look much like that in Figure 7. Clicking the right arrow again at the lower-right-hand corner will submit the job to the computer and return the user to the "Job Manager" but with the new job submitted. Once the job is complete, the Job Manager will report it as such, and the user can analyze both the raw and HTML outputs for the required information for the exercise. The GUI results are relatively easy for the students to understand especially for things like geometry optimizations where the final structure of the molecule is visually depicted and can be manipulated by the student. The 3-dimensional visuals greatly enhance the student's ability to conceptualize the molecule.

Beyond the description of the steps used in the WebMO interface, the example "Is C_3H^+ present in the Horsehead nebula?" exercise is a fully developed lab that we have already implemented in some of our classes. In this lab, students are asked if they can use quantum chemical techniques to aid in the detection of a new molecule in space. Pety and coworkers (*31*) observed rotational lines toward the Horsehead nebula and suggest C_3H^+ as the carrier of the lines. The students then perform geometry optimizations in order to derive the rotational constants of this molecule as well as several others including C_3H^-, $NNOH^+$, and HCCN that are related by mass to the test molecule. The students must compute rotational constants for the sample molecules and compare their values to those determined by Pety and coworkers. The students are then asked to discuss which molecules can be eliminated as carriers, which three are the most likely to be the carrier, and which one they believe actually is the carrier based on their data and understanding of rotational spectroscopy. The stated learning objectives for this lab indicate that the students will: 1) apply methods and basis sets for the study of rotational spectroscopy, 2) formulate and analyze data in order to evaluate previous scientific claims, and 3) propose and judge logical alternatives to previous scientific claims. It is hoped that the richness of this exercise and its application to astronomy will engage students not only in computational chemistry but serve to enhance their knowledge of spectroscopy as well. Each exercise listed on the PSI4Education website has the same rigor and depth creating a full set of active and collaborative learning experiences based on computational chemistry to further develop students' chemical intuition and understanding.

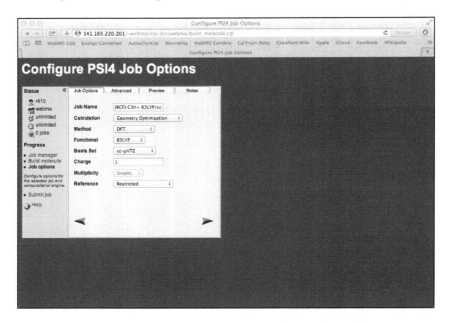

Figure 7. The "Configure Job" window. (Printed with permission from reference (19).)

Future Goals

PSI4Education is also open to new ideas and new contributions, especially for courses besides quantum chemistry. Labs related to organic, inorganic, thermodynamics/kinetics, and analytical chemistry are certainly possible, and we welcome contributions in these areas. We ask that new contributions be designed for the WebMO/PSI4 interface with detailed instructions and an answer worksheet formatted in the PSI4Education style along with an instructor guide. We expect that each lab submitted will have been tested by the author at their own institution. In this process, we expect that the author will include student feedback in refining the activity and will convey this information to the PSI4Education team as part of the submission. Then, the activity will be vetted by one or more of the core PSI4Education team and be tested by a student at a different institution. If the exercise is deemed satisfactory, it will subsequently be published in its respective classification on the PSI4Education website. The PSI4Education project is intended to be dynamic; updates and changes will take place on a regular basis as lab activities are further improved. We hope that more members contributing more labs will fully develop freely available computational chemistry resources with free software across the chemistry curriculum.

Since WebMO is relatively easy-to-use, we hope that PSI4Education resources can be utilized in the high school chemistry classroom, as well. One of us (RCF) has already taught a one-day summer course for gifted high school students where they utilized WebMO to design and test molecules for comparison of their various spectroscopic features to observed astronomical features.

As mentioned previously, commercial cloud computing (CCC) represents a new means of utilizing computational resources (*18*). One benefit to this environment is that a created disk image can be copied and used by any user with the proper granted access (*35*). Current members of the PSI4Education team and other collaborators are exploring how "plug-and-play" resources may be created such that an educator would actually never have to install any of the programs but would only have to manage their CCC account and necessary WebMO setup.

Conclusions

Computational chemistry has enhanced and is revolutionizing chemical research. As such, students must be taught these skills. PSI4Education exists to lower the barriers to the use of computational chemistry in the undergraduate classroom. By providing freely available and flexible instructional resources making use of free or low-cost computer software, it is hoped that the uptake of such materials will increase the use of computational chemistry as a teaching tool and provide modern students with modern skills upon graduation. Additionally, we are providing assistance and support in the installation and setup of these computational resources. We are open and welcoming to further growth of this team and to new ideas and resources being submitted in order to expand the reach of computational chemistry to the twenty-first century learner. As a final note, the WebMO/PSI4 setup described here may also be used for undergraduate research, potentially enhancing student learning outcomes even more. Such skills will make our students more competitive for postgraduate studies or industrial employment and will increase the quality of the chemistry education given.

Acknowledgments

The PSI4Education team members would like to thank the PSI4 team for hosting our web outlets on their website. We would also like to extend our appreciation to J. R. Schmidt and Bill Polik of WebMO for their assistance in getting PSI4 and WebMO linked and for their continued encouragement of this project. Thanks are also due to Russell F. Thackston of Georgia Southern University for his work in helping us to explore how PSI4Education can be utilized via cloud computing. RCF would like to acknowledge Georgia Southern

University for the provision of start-up funds necessary to begin this work. ARM acknowledges support from the Cal Poly Extramural Funding Initiative. CDS acknowledges support from the National Science Foundation (Grant No. CHE-1300497). Lastly, we would like to thank our current and future users of these resources for encouraging your students to learn new and cutting-edge computational chemistry skills.

References

1. Szabo, A.; Ostlund, N. S. *Modern Quantum Chemistry: Introduction to Advanced Electronic Structure Theory*; Dover: Mineola, NY, 1996.
2. Pilar, F. L. *Elementary Quantum Chemistry*, 2nd ed.; Dover: Mineola, NY, 1990.
3. Roothaan, C. C. J. *J. Molec. Struct.* **1991**, *234*, 1–12.
4. O'Grady, C. E.; Talpey, P.; Elgren, T. E.; Van Wynsberghe, A. W. *Annu. Rep. Comput. Chem.* **2014**, *10*, 168–187.
5. Schaefer, H. F., III *J. Molec. Struct.* **2001**, *573*, 129–137.
6. Lewars, E. G. *Computational Chemistry: Introduction to the Theory and Applications of Molecular and Quantum Mechanics*, 2nd ed.; Springer: Dordrecht, 2010.
7. Duke, B. J.; O'Leary, B. *J. Chem. Educ.* **1992**, *75*, 529–533.
8. Martin, N. H. *J. Chem. Educ.* **1998**, *75*, 241–243.
9. Shields, G. *J. Chem. Educ.* **1994**, *71*, 951–953.
10. Bryce, D. L.; Wasylishen, R. E. *J. Chem. Educ.* **2001**, *78*, 124–133.
11. Paselk, R. A.; Zoellner, R. W. *J. Chem. Educ.* **2002**, *79*, 1192–1195.
12. Pearson, J. K. *J. Chem. Educ.* **2007**, *84*, 1323–1325.
13. Simpson, S.; Lonie, D. C.; Chen, J.; Zurek, E. *J. Chem. Educ.* **2013**, *90*, 651–655.
14. Frey, E. R.; Sygula, A.; Hammer, N. I. *J. Chem. Educ.* **2014**, *91*, 2186–2190.
15. Nassabeh, N.; Tran, M.; Fleming, P. E. *J. Chem. Educ.* **2014**, *91*, 1248–1253.
16. Ochterski, J. W. *J. Chem. Educ.* **2014**, *91*, 817–822.
17. Wijtmans, M.; van Rens, L.; van Muijlwijk-Koezen, J. E. *J. Chem. Educ.* **2014**, *91*, 1830–1837.
18. Thackston, R. F.; Fortenberry, R. C. *J. Comput. Chem.* **2015**, *36*, 926–933.
19. Schmidt, J. R.; Polik, W. F. *WebMO Enterprise version 13.0*; WebMO, LLC: Holland, MI, 2013. http://www.webmo.net (accessed May 30, 2015).
20. Turney, J. M.; Simmonett, A. C.; Parrish, R. M.; Hohenstein, E. G.; Evangelista, F. A.; Fermann, J. T.; Mintz, B. J.; Burns, L. A.; Wilke, J. J.; Abrams, M. L.; Russ, N. J.; Leininger, M. L.; Janssen, C. L.; Seidl, E. T.; Allen, W. D.; Schaefer, H. F., III; King, R. A.; Valeev, E. F.; Sherrill, C. D.; Crawford, T. D. *WIREs Comput. Molec. Sci.* **2012**, *2* (4), 556–565.
21. Geng, Y.; Takatani, T.; Hohenstein, E. G.; Sherrill, C. D. *J. Phys. Chem. A* **2010**, *114*, 3576–3582.
22. Hohenstein, E. G.; Duan, J.; Sherrill, C. D. *J. Am. Chem. Soc.* **2011**, *133*, 13244–13247.

23. Fortenberry, R. C.; Morgan, W. J.; Enyard, J. D. *J. Phys. Chem. A* **2014**, *118*, 10763–10769.

24. Parker, T. M.; Burns, L. A.; Parrish, R. M.; Ryno, A. G.; Sherrill, C. D. *J. Chem. Phys.* **2014**, *140*, 94106.

25. Theis, R. A.; Morgan, W. J.; Fortenberry, R. C. *Mon. Not. R. Astron. Soc.* **2014**, *446*, 195–204.

26. Morgan, W. J.; Fortenberry, R. C. *Spectrochim. Acta, Part A.* **2015**, *135*, 965–972.

27. Foresman, J. B.; Frisch, Ae. *Exploring Chemistry with Electronic Structure Methods: A Guide to Using Gaussian*, 2nd ed.; Gaussian, Inc.: Pittsburgh, PA, 1996.

28. Pavia, D. L.; Kriz, G. S.; Lampman, G. M.; Engel, R. G. *A Microscale Approach to Organic Laboratory Techniques*, 5th ed.; Brooks/Cole: Belmont, CA, 2012.

29. Heine, T.; Joswig, J.-O.; Gelessus, A. *Computational Chemistry Workbook*; Weinheim, 2009.

30. Frisch, M. J.; Trucks, G. W.; Schlegel, H. B.; Scuseria, G. E.; Robb, M. A.; Cheeseman, J. R.; Montgomery, J. A.; Jr.; Vreven, T.; Kudin, K. N.; Burant, J. C.; Millam, J. M.; Iyengar, S. S.; Tomasi, J.; Barone, V.; Mennucci, B.; Cossi, M.; Scalmani, G.; Rega, N.; Petersson, G. A.; Nakatsuji, H.; Hada, M.; Ehara, M.; Toyota, K.; Fukuda, R.; Hasegawa, J.; Ishida, M.; Nakajima, T.; Honda, Y.; Kitao, O.; Nakai, H.; Klene, M.; Li, X.; Knox, J. E.; Hratchian, H. P.; Cross, J. B.; Adamo, C.; Jaramillo, J.; Gomperts, R.; Stratmann, R. E.; Yazyev, O.; Austin, A. J.; Cammi, R.; Pomelli, C.; Ochterski, J. W.; Ayala, P. Y.; Morokuma, K.; Voth, G. A.; Salvador, P.; Dannenberg, J. J.; Zakrzewski, V. G.; Dapprich, S.; Daniels, A. D.; Strain, M. C.; Farkas, O.; Malick, D. K.; Rabuck, A. D.; Raghavachari, K.; Foresman, J. B.; Ortiz, J. V; Cui, Q.; Baboul, A. G.; Clifford, S.; Cioslowski, J.; Stefanov, B. B.; Liu, G.; Liashenko, A.; Piskorz, P.; Komaromi, I.; Martin, R. L.; Fox, D. J.; Keith, T.; Al-Laham, M. A.; Peng, C. Y.; Nanayakkara, A.; Challacombe, M.; Gill, P. M. W.; Johnson, B.; Chen, W.; Wong, M. W.; Gonzalez, C.; Pople, J. A. *Gaussian-03*, 2003.

31. Pety, J.; Gratier, P.; Guzmán, V.; Roueff, E.; Gerin, M.; Goicoechea, J. R.; Bardeau, S.; Sievers, A.; Petit, F. Le; Bourlot, J. Le; Belloche, A.; Talbi, D. *Astron. Astrophys.* **2012**, *548*, A68.

32. Huang, X.; Fortenberry, R. C.; Lee, T. J. *Astrophys. J. Lett.* **2013**, *768*, 25.

33. Fortenberry, R. C.; Huang, X.; Crawford, T. D.; Lee, T. J. *Astrophys. J.* **2013**, *772*, 39.

34. Brünken, S.; Kluge, L.; Stoffels, A.; Asvany, O.; Schlemmer, S. *Astrophys. J.* **2014**, *783*, L4.

35. Wong, A. K. L.; Goscinski, A. M. *Future Gener. Comput. Syst.* **2013**, *29*, 1333–1344.

Curricular Endeavors

Chapter 8

The Power of Experiential Learning: Leveraging Your General Education Curriculum To Invigorate Your Chemistry Courses

Kimberlee Daus* and Rachel Rigsby

Department of Chemistry & Physics, Belmont University, 1900 Belmont Blvd., Nashville, Tennessee 37212, United States *E-mail: kim.daus@belmont.edu

In recent years, many institutions have undergone general education reform in attempts to improve student engagement and retention. The resulting curriculums, which focus on high-impact practices, are wonderful avenues through which chemistry departments can explore options for new classes or rejuvenate existing classes. Within General Education at Belmont, Chemistry faculty have created Learning Community Courses on both the majors and non-majors level. Additionally, we have extensively used experiential learning pedagogies such as service learning and problem-based learning to encourage students to delve more deeply into real-world problems in an upper-division chemistry elective (Medicinal Chemistry) and in courses for non-science majors. This chapter describes the pedagogies gleaned from the General Education curriculum, the chemistry courses developed using these pedagogies, and student learning and engagement obtained through these courses.

Introduction

Increasing student success and engagement in college chemistry is challenging at best. In addition to addressing the wide range of experiences and abilities that students bring into the classroom, faculty must look for creative ways to make the curriculum relevant and timely for the students. A wealth of knowledge on research-based practices for improving teaching and learning is available in the literature. A leading figure in research on high-impact practices, George Kuh, has noted (*1*):

> "[W]hen I am asked, what one thing we can do to enhance student engagement and increase student success? I now have an answer: make it possible for every student to participate in at least two high-impact activities during his or her undergraduate program, one in the first year, and one taken later in relation to the major field. The obvious choices for incoming students are first-year seminars, learning communities, and service learning."

Many science programs have not taken full advantage of high impact practices such as experiential learning or learning communities (*2*). In the Chemistry Department at Belmont, we have successfully leveraged our revised general education program to implement high-impact practices at multiple levels in our curriculum. This includes a re-design of both introductory and upper-division courses in the chemistry major as well as the modification of general education chemistry classes. We have successfully offered courses as Learning Communities and have utilized experiential learning to enhance several other classes. In this chapter we provide examples of specific course modifications implemented to invigorate our chemistry courses.

General Education Reform

In 2004, in the midst of a changing climate in higher education, Belmont University proposed a reformed general education program. The Belmont Experience: Learning for Life, or BELL Core, passed with the overwhelming support of the faculty. In addition to traditional content courses (fine arts, humanities, science, mathematics, social science, religion, and physical education), it contained several key components recognized as high-impact educational practices by organizations such as the Association of American Colleges and Universities (AAC&U) (*3*). These included a first-year seminar, learning community courses, and senior capstones. In addition, multiple courses identified as global studies and experiential learning were required of every graduate.

The BELL Core was developed as a vertical curriculum (Table 1) for students as they matured through four stages of learning (launching, intersecting, broadening, and reflecting).

Table 1. BELL Core Signature Requirements and Traditional Content Courses by Year

Year	Courses
1	First Year Seminar, Writing, Learning Community, content
2	Speech, content
3	Junior Cornerstone, Writing, content
4	Senior Capstone, content

Students would enroll in a broad, non-disciplinary First Year Seminar (*launching*), then move into thematic Learning Communities, courses designed to explore how two disciplines studied a particular issue relevant to both disciplines (*intersecting*). This combination of courses was designed to provide students with a framework on 'ways of knowing', or how knowledge can be gathered, evaluated, or interpreted. These introductory courses were followed in the third year with a Junior Cornerstone, a course designed around broad disciplinary content with a focus on collaborative learning and application of content to real-world problems (*broadening*). Finally, a culminating Senior Capstone seminar would allow students to *reflect* on their educational experience with a focus on what constituted a 'meaningful life' and the transition to life after college.

Experiential Learning

In addition to the vertical design of the General Education curriculum and the inclusion of key courses, the BELL Core included an expectation of experiential learning. In the course of their undergraduate career, students must take a minimum of two courses designated as experiential learning (EL). Also known as learning through experience, experiential learning is a process through which students develop knowledge, skills, and values from direct experiences outside a traditional academic setting. Experiential learning is not new; educational psychologist John Dewey noted its importance in 1938 by stating, "There is an intimate and necessary relation between the process of actual experience and education" (*4*). Dewey believed that experiential learning created a rich learning environment in which critical thinking and problem solving thrived. David Kolb furthered this idea through the development of an experiential model (Figure 1); this model outlines the key tenets needed to ensure the success of experiential learning.

Figure 1. The Kolb Experiential Model (5).

The Kolb cycle reflects the natural process by which experiential learning occurs; typically learning starts with the experience itself. The "doing" aspect engages students in a direct practice of their learning and, consequently, results in them taking ownership of their own learning. Through the next process, observation and reflection on the experience, students break down their experiences and make connections between the experience and their learning. In the third phase, insight, students formulate new understandings based on their experiences which they then test (action) in the fourth phase. The cycle is repeated through new experiences.

Typical experiential learning venues include internships, service learning, undergraduate research, study abroad, and other creative and professional work experiences. At Belmont, chemistry majors, depending on their program requirements, can attain one EL credit either as an internship or through undergraduate research. In 2011 a second option was offered for Chemistry majors through a Service-Learning course in Medicinal Chemistry.

Pedagogically, Service-Learning offers many advantages for both students and instructors; by relating course content to a service project, students more readily connect their learning to a real-world context. This connection results in deeper learning and stronger skill sets (6, 7). Moreover, through these experiences, students develop a strong sense of community which serves to connect them further to the instructor and to the institution as a whole (8).

Incorporating service-learning in chemistry (and many of the sciences) is challenging due to a number of factors, one of which is the time commitment required of students in laboratory-based classes (9). Yet, chemistry faculty wanting to capitalize on the pedagogical advantages of service-learning have sought way to include it in their courses. Short term projects are popular options; recent service-learning projects in chemistry include using biochemistry to design pamphlets for distribution in homeless shelters (10) and teaching an organic chemistry lab synthesis of azo dyes to high school students in underserved areas (11). However, these short term projects do not meet the requirements for a course to be designated as "experiential learning—service-learning" at Belmont. Service-learning classes require students to commit 8–20+ hours of service work during the semester (12). Full requirements include:

- On-site community involvement that is intentionally linked to the course content
- Preparation for, as well as reflections on, the service experience
- A minimum of eight hours of direct engagement with the community partner/project
- The service must be for the service of community engagement and learning, not primarily for the development of personal or job-related skills—and the experience must be mutually beneficial to the community partner/project.

Embedding Service-Learning in Medicinal Chemistry offered several advantages: (1) Medicinal Chemistry is an elective course with no lab and thus, affords the instructor more flexibility for including SL projects; (2) content learned in the course is relevant to the community and easily communicated; (3) societal issues such as drug addiction and misuse are easy to incorporate into the curriculum; and (4) the topic lent itself to multiple S-L projects.

In the Medicinal Chemistry course, two service-learning projects were developed in partnership with CADCAT (Community Anti-Drug Coalitions Across Tennessee) as part of their Inter-Collegiate Prevention Task Force. The two projects are described below.

- Prescription Drug Disposal Program ("Drug Take Back"): Students participating in this project worked with faculty and students from two other regional universities in executing a disposal program for unused household pharmaceuticals. The event was held in a local county and was coordinated through the sheriff's department. Students researched and developed pamphlets that were distributed to the public (at the drug take back) on the hazards of different prescription drugs including the impact on the environment as a result of improper disposal. During the take-back day, students distributed their pamphlets to the public and assisted in counting and cataloging of prescription and over-the-counter drugs. This drug documentation was then used as part of an assessment study to evaluate the effectiveness of take-back days in local communities.
- Collegiate Assessment and Environmental Scans: Through this project students examined the current Belmont University alcohol and drug prevention plan and compared its alignment with Nashville's coalition plan to address underage drinking. One effort of CADCAT was to determine the relationship between advertisement and subsequent sales of alcohol- and nicotine-related products. Students conducted environmental scans of various Nashville convenience stores (through Nashville Prevention Partnerships, NPP); three teams of students evaluated 41 locations in the metropolitan and surrounding areas. The environmental scans involved an analysis of the signage posted on the exterior and interior of the sites with regard to the advertisement of alcohol. In addition to presenting their results to NPP and the class, students participating in this project also researched the chemistry of alcohol and nicotine.

During their project, students maintained reflective journals. A series of reflective prompts throughout the semester helped students to make connections between their service experiences and their learning. To provide deeper connections, prompts were intentionally linked to the service projects. For example, in the second prompt, students involved in the drug take-back were asked to talk with three people (of different ages) about practices in their homes concerning prescription drug disposal whereas students involved in the NPP project were asked to interview underage peers about factors that influenced their practices (of consumption of alcohol and nicotine). At the conclusion of the project, the majority of the students reported that the Service-Learning venture enhanced their learning in Medicinal Chemistry and helped them to gain an awareness of societal issues associated with drugs and alcohol abuse and misuse.

Learning Communities

In addition to the experiential learning requirements, Learning Communities were included in Belmont's reformed general education curriculum. Numerous institutions across the country have implemented Learning Communities; types of LCs vary and include paired courses with concurrent student enrollment, living-learning communities, and even faculty LCs whose members are interested in improving teaching and learning in their classrooms (*13, 14*). Student learning communities early in the college experience have been shown to promote deep learning and general gains among students. Initial research on the effectiveness of LCs concluded that participation in a learning community was associated with increases in student retention and academic performance (*15*). An extensive literature review on the efficacy of LCs can be found elsewhere (*16*).

LCs have been utilized by institutions in a variety of ways in the sciences. Some programs are using them to provide additional coursework support and faculty mentoring to a cohort of students (*17*); others, such as the University of Kentucky, allow students to enter STEM living-learning communities (*18*). However, these models do not fit the type of LCs at Belmont as they do not typically involve specific courses in the chemistry curriculum.

LCs in our curriculum are defined as two general education classes with concurrent enrollment; paired courses must come from two different content areas (science, social science, mathematics, humanities, fine arts, religion, or physical education). Paired classes are not team-taught; rather, each faculty person teaches his or her disciplinary course as a stand-alone course. Paired courses focus on an area of overlap in content or a common issue relevant to both disciplines. Faculty are encouraged to attend one another's classes when possible, and they may assign common readings or experiences, but faculty only teach in their respective disciplines. Learning outcomes include the development of students' ability to:

- distinguish between the kinds of knowledge and the types of thinking and learning processes that are represented in two disciplines
- recognize the interconnectedness of knowledge through examination of an area of overlap between two courses
- integrate learning from each of the disciplines into the other and provide a specific example of how something learned in one class contributed to understanding of the other
- evaluate various information and experiences from the perspective of each of the disciplines

An additional Learning Community expectation stemming from Belmont's recent institutional Quality Enhancement Plan is that all linked courses include a significant shared experiential component such as a field trip or service project where students encounter the disciplines being studied in the real world (*19*).

This experiential component is designed to address and reinforce students' sense of competency in the subjects studied through the experience and application of learning. For example, a theater and literature LC has attended a production of a play read in class followed by an in-depth interview with the cast and producer after the production; a biology and writing LC has visited an area park and written about their experiences from a scientific perspective.

Chemistry Learning Communities have been formed by linking chemistry with courses in the social sciences, humanities, and fine arts. Our department has taken advantage of the fact that students enroll in LCs in semester two of their academic career, when chemistry majors are enrolled in the second semester of General Chemistry. Accordingly, we have modified the second semester of our General Chemistry sequence and have offered it as part of several different Learning Communities. In one long-standing LC, General Chemistry 2 was paired with a political science public policy course. Faculty delivered key area content with a course theme of how chemists and policy makers view environmental issues. The culminating experience was a group project requiring students to collect and analyze an environmental water sample. An extensive laboratory analysis included typical freshman techniques such as titrations, construction and use of a calibration curve, and use of a simple spectrophotometer. After completing their multi-week study, student groups wrote policy papers and led presentations based on their findings informing mock legislators (their classmates) of their results in support of potential legislation concerning water quality in Tennessee.

We have also linked General Chemistry 2 with an Art Experience course; these courses focused on visual representations in the disciplines. Students finished the courses by creating models and artistic renderings of various molecules. More recently, we have offered General Chemistry linked with English literature with a course theme of forensics. Here faculty have used a thin-layer chromatography forensics activity to demonstrate molecular polarity and intermolecular forces, traditional general chemistry content.

It is important to note here that General Chemistry LCs follow the same course syllabus as non-LC sections, with very few modifications. Typically a handful of lectures throughout the semester are used to discuss common readings or show films in the learning community; non-LC sections use these days to spend additional time on topics of the professor's choosing, or to allow more flexibility in the schedule. Additionally, all sections follow the same lab schedule, which includes typical freshman labs as well as experiments implemented specifically for students in the LC. For example, in semesters when the chemistry/policy LC is offered, all students complete a series of water analysis labs using an environmental water sample. These labs were specifically designed to incorporate labs skills taught in traditional labs and include:

- *Determination of phosphate concentration*: Students collect visible absorbance data and construct a calibration curve from standards followed by analysis of the unknown.

- *Determination of metal ion concentration*: Students perform EDTA titrations of water samples; this lab includes the idea of repeating data in triplicate and error analysis.
- *Determination of water hardness, pH, etc*: Students use a commercial water analysis kit and interpret results based on provided keys.

We have also used the Learning Community model to increase the appeal of chemistry courses for non-science majors. Belmont has a significant undergraduate population interested in music, so we developed a learning community of chemistry and Popular Music History, a course that can fulfill Belmont's fine arts requirement. Instead of an area of overlap in course content, faculty focused on how practitioners of both disciplines affected, and were affected by, 20th century American culture. The chemistry course covered content typical for this type of class and used the American Chemical Society's '*Context in Chemistry*' text (*20*). Faculty teaching the courses used a variety of methods to help students see the importance of both disciplines. In one example, students viewed portions of *Doctor Atomic*, a contemporary opera displaying the stress and emotion leading up to the testing of the first atomic bomb. They then wrote reflections on both the musical and scientific elements represented in the production. Additionally, examples were chosen from the chemistry text and from a variety of current sources such as *Chemical & Engineering News* to reinforce the relevance of chemistry in society. As a final experience, student groups presented research projects on musical and scientific influences in American Culture from each decade in the 20th century. Topics included birth control, chemical warfare in Vietnam, and the polio vaccine.

In a second example, chemistry for non-science majors was linked with Economic Inquiry, a general education social science course. Students enrolled in the courses developed an understanding of fundamentals of chemistry and basic principles of economics while focusing on overlapping areas of content such as the economics and chemistry of food, the pharmaceutical industry, and energy production. This LC also included an experiential component which was a class visit to a non-operational nuclear power facility. Through a pre-tour lecture, hands-on demonstrations, and comments throughout a 2-hour tour, students were able to see a practical application of nuclear chemistry as well as various principles of economics in action. Learning about detailed specifications, down to the exact types of bolt used and regulations on how bolts were to be tightened in the facility as well as how much care was taken to pour each section of concrete in the cooling towers helped students understand the costs involved in building and maintaining a nuclear facility. The primary coolant loop, turbine, and spent fuel rod storage pools helped students see their textbook knowledge of chemistry in the real world. The courses culminated in student groups presenting research on the overlap of chemistry and economics in areas such as organic food production and marketing and the solar power industry.

Chemistry for non-science majors has also been linked with literature with a focus on forensic science. A second link with an art experience course artistically interpreted each element to create a 4 x 10 foot periodic table. Options for future LCs include a biochemistry- and nutritionally-themed chemistry course paired with wellness (physical education) and a technology-focused course paired with science fiction literature.

Junior Cornerstone Seminar

In the third year, Belmont students encounter their next course as part of Belmont's vertical curriculum, the Junior Cornerstone Seminar (JCS). JCS courses offer faculty yet another way to engage students in their discipline. Four principles guide faculty in the development of a disciplinary-based JCS at Belmont University:

- All JCS courses are experiential in nature and, thus, carry an EL designation. Faculty typically develop these courses as study abroad, undergraduate research, community-based research, or service-learning courses.
- JCS courses focus on real problems in the discipline and invite non-majors to grapple with current issues in the field.
- Collaborative work is an integral part of the JCS experience; problem-solving skills are honed through working problems in groups.
- Students further experience the discipline by communicating the results of their work through a disciplinary-appropriate venue.

Although there are different models in which teamwork and problem-solving can be incorporated into a JCS, many JCS faculty utilize Problem-Based Learning (PBL). The Problem-Based Learning pedagogy has its origins in medical schools; McMaster University (Canada) first adopted PBL over thirty years ago in an effort to improve student learning and information retention (21). Since then, PBL has been adopted broadly throughout all levels of education, and is seen in K-12 as one way to address the new standards (Common Core State Standards and New Generation Science Standards) (22). Pedagogically, PBL is the epitome of high-impact practices (HIP) in education (1, 23). At the core of this pedagogy are the problems themselves; problems are created based on relevant "real-world" scenarios and serve as the drivers for learning in PBL. Problems are intentionally complex and ambiguous such that students have to use higher level critical thinking skills (analysis, synthesis, evaluation, and creation) to resolve them. In working the problem, each group member is responsible for researching certain topics and, thus, becoming the "expert" in that field. All members share, evaluate, and expound on the groups' research. In addition to providing a meaningful context in which deeper learning occurs, the PBL process also provides an environment in which students further learn the fine art of collaboration.

Elsewhere, PBL has been used with good results in courses for Chemistry majors (24, 25); at Belmont we have also found that using Problem-Based Learning in non-majors courses is particularly attractive as the courses can be developed on topics that non-majors find of interest. Courses developed by our department as Junior Cornerstone Seminars include "Your Brain on Drugs: The Chemistry of Drugs and Addiction" and "Better Eating through Chemistry: The

Chemistry of Food and Cooking." As the latter course has been described in a previous volume (26), this chapter will explore the Chemistry of Drugs and Addiction course and will examine the integration of PBL and EL into the course.

"Your Brain on Drugs" was first offered in the spring of 2006 as part of the new BELL Core curriculum. The course was designed around three problems; each problem guided students in their research of the chemistry involved in drugs and addiction and the associated societal problems. Problems were developed such that students started with a broad topic and, through research and discussion of the chemistry involved, reached a problem resolution. Topics for the three problems are noted below along with chemistry expectations for the problems.

1) Was a suitemate under the influence of a drug? A new suitemate, Bobby, is acting strangely and you want to determine if his behavior is drug-related. Students researched various club drugs and, as they gained an understanding of drug-receptor binding, learned about atoms, elements, and compounds as well as bonding and intermolecular attractive forces.

2) How to deal with chronic pain? Following a car accident, a close friend, Abby, struggles with dealing with pain. Students explored different pain medications and treatments to develop a healthy alternative for Abby. In this problem, students learned about the connection between structure and function of the anti-inflammatory and pain medications in addition to states of matter.

3) Is the new law restricting sale of cold and allergy medications effective in decreasing the production of methamphetamine? Students researched the synthesis of methamphetamine, examined the possibility of restricting other starting materials than pseudoephedrine, and investigated the effectiveness of similar laws addressing methamphetamine synthesis in other states.

During most classes, students conducted group sessions while faculty provided guidance in research directions. In the group sessions, students presented their research, discussed and analyzed the group's research for that day, developed new areas of research, and assigned research areas for the next class meeting. Group assessment took place every two weeks. In the group assessment, members gave face-to-face feedback to each other on three aspects of collaboration: knowledge and self-directed learning, reasoning, and group and interpersonal skills. These open conversations served to model good practices in business communication.

Service-learning was selected as a natural EL component for this course. In order for students to more fully appreciate the human side of addiction, the class partnered with the Magdalene House, "a residential program for women who have survived lives of prostitution, trafficking, addiction, and life on the streets (27)." Students became acquainted with Magdalene women through group and NA/AA meetings and developed service projects that enhanced the daily lives of the residents. The service learning experience culminated in a dinner that was planned and prepared by the class for the women. Throughout the semester students reflected on their experiences through writing activities, online and in-class discussions, and authoring letters to incoming JCS students.

Chemistry Junior Cornerstone Seminar courses have proven to be very attractive to students and are some of the first to fill during registration. In addition to interest in the topics, students find learning through PBL to be invigorating and relevant. As one student in the Drug and Addiction JCS noted, "After taking this class, I would consider myself an advocate of PBL. It has given me back my belief in education to the point that I wish I would have taken a class like this several years ago." Another student commented, "(PBL)…also increased my research skills…and made me dig to a deeper level than I may have reached in a lecture-style class."

As mentioned previously, another version of JCS offered at Belmont is the Chemistry of Food and Cooking. In this course, particularly, we find that the topical nature invites deeper learning. Student comments at the end of the semester reflect the connections they made between chemistry content, cooking, and their own lives:

- *These 3 cooking days....have helped me to understand more about the science of food.*
- *My perception of science has changed....I find myself looking for chemistry in places other than in the classroom.*
- *Learning the fundamentals of chemistry was great, but being able to actually apply them to understanding how oils react, how protein can be substituted, and other specific factors was unique and interesting.*

Conclusions

Chemistry faculty at Belmont University have taken extensive advantage of the exciting opportunities provided by our reformed general education curriculum. Not only have we found the opportunities to teach Learning Community, Junior Cornerstone, and Experiential Learning courses for both chemistry majors and non-majors to be deeply rewarding and invigorating, we find that the biggest payoff is with the students; overall, students taking these courses are much more engaged, experience deeper learning, and form stronger communities within the classes. Chemistry faculty continue to explore new ways to leverage Belmont's general education program to improve teaching and learning in our classrooms.

References

1. Kuh, G. D., Schneider, C. G., & Association of American Colleges and Universities. *High-Impact Educational Practices: What They Are, Who Has Access to Them, and Why They Matter*; Association of American Colleges and Universities: Washington, DC, 2008.
2. Driscoll, W. d; Gelabert, M.; Richardson, N. *J. Chem. Educ.* **2010**, *87*, 49–53.

3. High-Impact Practices. http://www.aacu.org/resources/high-impact-practices (accessed December 14, 2014).

4. Dewey, J. *Logic, the Theory of Inquiry*; H. Holt and Company: New York, 1938.

5. Learning-Theories.com. http://www.learning-theories.com/experiential-learning-kolb.html (accessed January 22, 2015)

6. Zlotkowski, E. Introduction. In *Service-Learning and the First-Year Experience: Preparing Students for Personal Success and Civic Responsibility*; Monograph 34; Zlotkowski, E., Ed.; University of South Carolina, National Resource Center for the First-Year Experience and Students in Transition: Columbia, SC, 2002.

7. Esson, J. M.; Stevens-Truss, R.; Thomas, A. *J. Chem. Educ.* **2005**, *82*, 1168–1173.

8. Vogelgesang, L. J.; Ikeda, E. K.; Gilmartin, S. K.; Keup, J. R. Service-Learning and the First-Year Experience: Outcomes Related to Learning and Persistence. In *Service-Learning and the First-Year Experience: Preparing Students for Personal Success and Civic Responsibility*; Monograph 34; Zlotkowski, E., Ed.; University of South Carolina, National Resource Center for the First-Year Experience and Students in Transition: Columbia, SC, 2002; pp 15–24.

9. Sutheimer, S. *J. Chem. Educ.* **2008**, *85*, 231–233.

10. Harrison, M.A.; Dunbar, D.; Lopatto, L. *J. Chem. Educ.* **2013**, *90*, 210–214.

11. Glover, S.R.; Sewry, J.D.; Bromley, C.L.; Davies-Coleman, M.T.; Hlengwa, A. *J. Chem. Ed.* **2013**, *90*, 578–583.

12. The BELL Core. http://www.belmont.edu/bellcore/pdf/Experiential%20Learning%20Application.pdf (accessed January 22, 2015).

13. Zhao, C-M.; Kuh, G. D. *Research in Higher Education* **2004**, 115–138.

14. What Is a Faculty and Professional Learning Community? http://www.units.miamioh.edu/flc/whatis.php (accessed January 5, 2015).

15. Taylor, K. with Moore, W. S., MacGregor, J., Lindblad, J. *Learning Community Research and Assessment: What We Know Now. National Learning Communities Project Monograph Series*; The Evergreen State College, Washington Center for Improving the Quality of Undergraduate Education: Olympia, WA, 2003.

16. Brownell, J. E.; Swaner, L. E. *Five High-Impact Practices: Research on Learning Outcomes, Completion and Quality*; Association of American Colleges and Universities: Washington, DC, 2010.

17. Chemistry learning community. http://chem.ou.edu/chemistry-learning-community (accessed January 22, 2015).

18. STEMCats Program. https://stemcats.as.uky.edu/stemcats-program (accessed January 22, 2015).

19. Hunter, M. S. *Helping Sophomores Succeed: Understanding and Improving the Second-Year Experience*; Jossey-Bass: San Francisco, 2010.

20. American Chemical Society. *Chemistry in Context: Applying Chemistry to Society*, 7th ed.; McGraw-Hill; New York, 2012.

21. Background of Problem-Based Learning. http://www.samford.edu/ctls/archives.aspx?id=2147484113 (accessed December 29, 2014).

22. A Natural Fit—PBL, STEM and Technology Integration. http://www.k12blueprint.com/content/natural-fit-pbl-stem-and-technology-integration (accessed December 29, 2014).

23. What Is Problem-Based Learning? http://www.pbl.uci.edu/whatispbl.html (accessed December 16, 2014).

24. Dods, R. F. *J. Chem. Educ.* **1996**, *73*, 225228.

25. Ram, P. *J. Chem. Educ.* **1999**, *76*, 1122–1126.

26. Daus, K. A. Better Eating through Chemistry: Using Chemistry to Explore and Improve Local Cuisine. In *Using Food to Stimulate Interest in the Chemistry Classroom*; Symcox, K. Ed; ACS Symposium Series 1130; American Chemical Society: Washington, DC, 2013; pp 11–21.

27. About Magdalene. http://www.thistlefarms.org/index.php/about-magdalene (accessed January 10, 2015).

Chapter 9

The Green Chemistry Commitment

Transforming Chemistry Education in Higher Education

Amy S. Cannon*,[1] and Irvin J. Levy[2]

[1]Executive Director, Beyond Benign, 100 Research Drive,
Wilmington, Massachusetts 01887, United States
[2]Department of Chemistry, Gordon College, 255 Grapevine Road,
Wenham, Massachusetts 01984, United States
*E-mail: amy_cannon@beyondbenign.org

The Green Chemistry Commitment (GCC) is currently the
only nationwide program specifically designed to encourage,
empower, and celebrate entire departments of chemistry that
transform their curriculum through green chemistry. The GCC
challenges departments to ensure that each of their graduates is
trained in four pillars of green chemistry: Theory, Toxicology,
Laboratory Skills, and Application. The GCC is a distinctive
for an institution yet neither exclusive nor prescriptive. This
chapter presents case studies showcasing how two very different
chemistry departments (one a small undergraduate college; the
other a major research university) have implemented the GCC
at their institutions.

Introduction

Chemistry students are uniquely important as we consider the next
generation and its impact on the environment. A 2014 survey of 1,821 adults
by the Pew Research Center found that 32% of Millenials describe themselves
as environmentalists (1). Interestingly, concern for the environment has
become increasingly important during the childhood of these young adults
(2). Consequently, it is disconcerting that only one-third of those children *now*
self-describe as an environmentalist.

Green chemistry provides a scaffold that can inspire young people to produce
and utilize technologies that have been developed to be inherently safer for human

health and the environment. The appearance of multiple editions of chemistry texts that include green chemistry and sustainability during the same period as the Pew Research Center study indicates that faculty are actively pursuing these topics with their students (3, 4).

While individual faculty have incorporated green chemistry into their individual classes, laboratories, and research agendas in the past two decades (5), more needs to be done to assure that green chemistry education becomes part of the training of all chemistry students. The attitudinal surveys indicate that a majority of students will not seek this training on their own; consequently, the Green Chemistry Commitment (GCC) provides an institutional framework to encourage this training (6). The GCC is distinctive for an institution yet neither exclusive nor prescriptive.

Green Chemistry

Traditional methods for addressing pollution have included mitigation controls and end-of-pipe technologies. Since the early 1990's, it has been realized that these technologies are not sufficient to prevent pollution and the release of hazardous chemicals. There has been a clear shift towards preventative technologies that address pollution at the beginning, design stages of a product life cycle. Green chemistry has been recognized as a key aspect to these new approaches and has proven to be central to the development of materials and products that have reduced hazard and pollution, increased energy efficiency, as well as numerous other benefits. Green chemistry is being adopted widely throughout the chemical industry as a means for cost savings and enhancing efficiency. In a 2011 report from Pike Research (7), it was reported that the green chemistry industry will become a 100 billion dollar sector by 2020, with more than $20 billion of the growth in the U.S. The use of green chemistry will save the chemical industry more than $65.5 billion by 2020. In order to support this shift in the chemical industry and advance green chemistry throughout the U.S. and internationally a change must occur in how we are training current and next generation scientists.

Green Chemistry has been clearly defined since 1998, when Paul Anastas and John Warner published the book *Green Chemistry: Theory and Practice* (8). Throughout the years, green chemistry has gained much attention as a research framework, a business opportunity and a teaching tool. Green Chemistry is the design of chemical products or processes that reduce or eliminate the use or generation of hazardous substances. Green Chemistry principles have been adopted by researchers and educators throughout academia, industry and government. Despite the wide adoption of green chemistry principles, there remains a key missing piece to a chemist's education, that of understanding molecular hazards and toxicology. For Green Chemistry to be successfully integrated into research programs, both academic and industrial, the scientists must have a mechanistic understanding of how chemicals impact human health and the environment (9). Through this mechanistic understanding, scientists can design molecules that have reduced hazards to human health and the environment and ecosystem, an approach that is the best method for pollution prevention and avoiding the use and generation of hazardous chemicals.

In the academic year 2008–2009, U.S. colleges and universities that offered a chemistry degree approved by the American Chemical Society (ACS) granted 14,577 bachelor's degrees in chemistry, 1,986 master's degrees, and 2,543 doctoral degrees. Thus more than 19,000 students were trained in chemistry in the United States in just one year. More than 600 colleges and universities offer ACS-approved degree programs in chemistry (*10*). Only one of these programs requires classes in toxicology or environmental impacts: the University of Massachusetts Boston's Ph.D. program, from which two Ph.D. students graduated in the academic year of 2008–2009 (*11*, *12*). However, *not one undergraduate institution* requires courses in toxicology or environmental impacts. Learning about how to identify and avoid using or making toxic materials is essentially absent from the education of chemists today. This is a key aspect to green chemistry education.

Green Chemistry Education in the United States

Chemistry faculty have been incorporating green chemistry into chemistry courses for a number of years and the adoption of green chemistry has been on the rise since 1990. The early adoption of green chemistry has focused on reducing the use of hazardous chemicals in laboratory courses such as organic chemistry and has been catalyzed by the development of numerous resources and professional development opportunities, such as the Greener Educational Materials Database (GEMs) (*13*) and the Green Chemistry in Education Workshop, hosted by the University of Oregon (*14*). Tremendous advances have come through the development of greener and safer laboratory experiments that demonstrate alternative means for performing reactions under safer and less hazardous conditions. The obvious benefits include environmental, health and safety improvements due to the reduction in the use of hazardous chemicals in the teaching laboratory. However, the benefits include economic advantages as well as savings in the form of reduction of hazardous waste. For example, in a three-year pilot study at St. Olaf College, green chemistry laboratory experiments were implemented and the synthesis laboratory course saw a 30% decrease in hazardous waste (*15*). Also, new case studies show that by implementing one green chemistry experiment within a course such as organic chemistry, savings in purchasing and waste disposal costs can be realized (*16*).

In the field of green chemistry today, there is a movement towards teaching toxicology concepts to chemistry students so that they have a mechanistic understanding of how chemicals impact humans and the environment. Some institutions have begun efforts to create their own courses on toxicology on their campuses. The department of chemistry at the University of California, Berkeley has begun a seminar series and developed a one-unit class for their chemistry graduate students to introduce these topics (*17*). South Dakota State University chemistry faculty have created a new toxicology course for their majors (*18*), Simmons College runs a toxicology course for their chemistry majors (*19*),

Gordon College incorporates toxicology into the junior/senior seminar program for all undergraduate majors (20), Grand Valley State University has also used webinars as a means to introduce topics of toxicology (21), and many other institutions are beginning to teach toxicology concepts within their programs and courses. Also, many more institutions have shown interest, but do not have the resources or knowledgebase to implement a toxicology course for chemistry majors on their campus. Through collaborations, more resources in this area are being developed which will further enable faculty to adopt toxicology and related topics in their courses.

The Green Chemistry Commitment

The Green Chemistry Commitment (GCC) (22) is inspired by other successful programs that have adopted non-regulatory approaches to change, such as the American College and University Presidents' Climate Commitment, organized by Second Nature (23). The Presidents' Climate Commitment is an institutional commitment that is taken on by Presidents of colleges and universities to bring their campus to climate neutrality. The approach is non-regulatory and is a means for campuses to get involved with solving the problem of global climate change. The GCC was inspired by this program due to the pro-active approach to solving global problems, although the GCC uses a different approach.

The GCC seeks to build on the efforts of leaders in the field of green chemistry to systemically change chemistry education to reflect the changes in industry. The GCC is shaped and led by an advisory board currently comprised of faculty members from chemistry departments across the United States, representing large and small academic institutions. The advisory board helped to shape the GCC and provide direction for developing Green Chemistry Student Learning Objectives. The student learning objectives were found to be a way of focusing on what 21st century chemistry education should look like. With the recognition that implementing the green chemistry student learning objectives will be unique at each academic institution, this format provides flexibility within a realistic framework for guiding change in academia.

When an academic institution signs on to the GCC, they agree that upon graduation, chemistry majors should have proficiency in the following essential green chemistry competencies:

- **Theory**: Have a working knowledge of the Twelve Principles of Green Chemistry (8)
- **Toxicology**: Have an understanding of the basic principles of toxicology, the molecular mechanisms of how chemicals affect human health and the environment, and the resources to identify and assess molecular hazards
- **Laboratory Skills**: Possess the ability to assess chemical products and processes and design greener alternatives when appropriate
- **Application**: Be prepared to serve society in their professional capacity as scientists and professionals through the articulation, evaluation and

employment of methods and chemicals that are benign for human health and the environment

One of the four primary student-learning objectives states, "Students will have an understanding of the principles of toxicology, the molecular mechanisms of how chemicals affect human health and the environment, and the resources to identify and assess molecular hazards." As signers to the GCC, these institutions are looking to incorporate toxicology and environmental impact in to their undergraduate and graduate courses. Of the four green chemistry student learning objectives, this one presents the greatest challenge to faculty and departments.

Why Toxicology? Chemists are molecular designers and work with the fundamental building blocks that make up our industrial society. As chemists work towards greener, safer means for creating these building blocks, our society can realize tremendous benefits through the reduction in the use and generation of hazardous substances. As molecular designers, chemists require knowledge about how chemical structures and properties influence toxicity and environmental impact. Currently this knowledge is absent from the training of a chemist. By giving chemistry students a better understanding of how chemicals impact human health and the environment, these students can be better prepared to design greener, safer chemical products and processes.

The challenges of implementing the toxicology student learning objective are complex: faculty members typically are not trained in toxicology concepts and may lack the expertise; departments often do not have the resources to develop and teach an additional course; there is a lack of curriculum resources to teach toxicology concepts; along with many additional challenges to introducing a new topic area into an already jam-packed chemistry curriculum. Despite the challenges, there are many institutions beginning to teach these concepts, as mentioned previously (*17–21*). There are also new resources and professional development opportunities for faculty that are either under development or already offered to the community (*24, 25*). There are still many obstacles to including toxicology concepts within chemistry courses and programs, but the path towards including these topics is becoming more clear as more resources are becoming available.

Why the Green Chemistry Commitment?

The GCC offers an opportunity for academic institutions to unite around common goals. Through a collective voice, the GCC's signing institutions can help to inspire other institutions to get involved with green chemistry and transform their own institutions. Together, signing institutions of the GCC can also help to influence other initiatives that affect academia, such as funding agencies, degree program certifying institutions, and other govern-mental and non-governmental organizations.

The GCC tracks progress to implementing green chemistry in academia using a streamlined reporting process. Departments track past accomplishments and map out future goals. The accomplishments of participating institutions are then highlighted through illustrations and case studies to inspire others to adopt green chemistry practices.

The GCC also works collaboratively with multiple institutions through working groups that are comprised of faculty members from both signing and non-signing institutions.

Some Approaches for Adopting the GCC

At this writing over two dozen chemistry departments have joined the GCC. These institutions are in various regions of the country and the institutions themselves vary considerably. The implementation of the green chemistry student learning objectives can look very different from institution to institution due to the differences in curriculur mapping at those schools. However, the two case studies presented within this chapter can offer some guidance or inspiration for small and large academic institutions looking to adopt green chemistry at their own institution. Faculty at GCC member institutions are also available to discuss curriculum mapping with other departments that aspire to join the GCC. Interested faculty can find specific contact information at the GCC website (22).

In this section we will present case studies from two very different institutions: Gordon College, a small private liberal arts college and University of California, Berkeley, a very high research activity public university.

Green Chemistry at Gordon College

Gordon College is a small faith-based liberal arts college in the greater Boston region. The chemistry department has approximately 40–50 majors in several concentrations (Professional, Biochemistry, Health Professions, Secondary Education). Faculty and students at Gordon College became involved in green chemistry in the mid-2000's, largely because of the interest of students rather than faculty (26, 27).

Students are introduced to the "Theory" elements of the GCC throughout their coursework at Gordon College. For example, green chemistry metrics of atom economy (28) and E-factor (29) are presented in the first year curriculum alongside traditional metrics such as percentage yield in a reaction. During the Organic Chemistry year students are required to become competent in their understanding of all 12 of the Princples of Green Chemistry. These second year students are required to develop a significant outreach activity that decribes the concepts of green chemistry to an external audience. This form of service learning requires the students to develop a level of expertise that would not be mandatory for a quiz or test. It is interesting to note that the non-chemistry majors in the courses (the majority of students) find this major assignment to be especially compelling, rather than an extra burden (30).

The "Toxicology" pillar of the GCC is handled at Gordon College through guided readings and lectures on toxicology that are now part of the Junior/Senior seminar program. Students at Gordon College are required to participate in the seminar program for four semesters. The major themes of the seminar rotates on a two-year cycle. One of the four themes has been toxicology for the past several cycles of the series. Additionally, students are introduced to toxicology through a laboratory experiment in the organic chemistry sequence. In this experiment, created by students at Gordon College, the effect on germination and root elongation of lettuce seeds is used as an indicator of terrestrial ecotoxicity of several common organic substances (31).

The "Laboratory Skills" at Gordon College require students to become familiar with methods to find information about the substances that they are handling. For example, students in Organic Chemistry are required to compile GHS safety information about all substances prior to their use in the laboratory in the same way that they were once required to merely determine molecular weight and physical properties. Students entering the lab can reliably speak about the hazards of the substances that they will use on a given day. Such information can be acquired from the Safety Data Sheets (SDS) on file or through online sources (32). Students working on research projects are challenged to find safer solvents and conditions for their processes using various sources such as Sigma Aldrich's Greener Solvent Alternatives (33).

The "Applications" pillar of the GCC is the most lofty. At Gordon College our Green Chemistry outreach program is one way that we begin to train students to become chemists engaged in a lifetime of service to their comm-unities, broadly imagined. Longitudinal study will be valuable to determine whether these lessons truly have helped to develop chemists who will practice chemistry in ways that are inherently safer for human health and the environment. One anecdotal event that gives us cause for hope was described by a professor leading a group of physical chemistry researchers:

"My environmental/materials science research group is not inherently involved in green chemistry, but one of my students who is deeply steeped in the green chemistry culture is building an offshoot project within my group that is a greener approach to the main focus of our group right now: the optimization of polymeric materials as photocatalytic support (34)."

Green Chemistry at the University of California, Berkeley

The College of Chemistry at the University of California, Berkeley is the #1 top-ranked school for chemistry according to U.S. News & World Report (35). The College has a large undergraduate program, graduating roughly 40 students per year in chemistry, 60 students in chemical biology, and 130 in chemical engineering. The graduate program awards 70–80 Ph.D. degrees annually in chemistry. With a large program and a large number of students, the college has taken on a different approach to implementing green chemistry. The green chemistry education at UC Berkeley has been catalyzed by the development of an interdisciplinary center that spans multiple colleges and disciplines, the Berkeley Center for Green Chemistry (BCGC). Founded in 2010, the BCGC uses an interdisciplinary approach to green chemistry education, bringing together faculty and students from chemistry, public health, law, business, and environmental science. The BCGC has led change in the chemistry curriculum at UC Berkeley, which has begun with the transforming of the introductory chemistry course and has also involved the development of many interdisciplinary courses and curricula materials (*36*).

The "Theory" and "Laboratory Skills" elements of the GCC are carried out at UC Berkeley through the introduction of green chemistry into new laboratory experiments that are used in the introductory chemistry course, which is taken by 2,500 students annually. The laboratory experiments explore large technological challenges that face our society today through a multidisciplinary lense. The laboratory experiments include topics such as biofuels, sustainable polymers, and acids in the environment (*36*). Students are introduced to sustainability and green chemistry topics throughout the semester and are challenged to address global problems.

The "Toxicology" pillar of the GCC is carried out in the undergraduate curriculum by weaving concepts within the introductory courses, such as the measuring of octanol-water partition coefficients during an introductory chemistry laboratory experiment, an endpoint that can predict bioaccumulation in the environment. The BCGC also offers graduate courses that delve deeper into toxicology topics, such as a course entitled "The Basics of Toxicology for Green Molecular Design" that focused on understanding the basic principles of toxicology, understanding modes of action, and how to use tools and metrics to evaluate the hazard profile of chemical substances (*17*).

The BCGC continues to develop new laboratory experiments and implement them into their additional undergraduate courses, while new graduate courses have been developed such as an ethics course that explores the challenges and ethical considerations of implementing greener chemistry in our society (*37*).

A Call to Arms

"A sustainable world is one where people can escape poverty and enjoy decent work without harming the earth's essential ecosystems and resources; where people can stay healthy and get the food and water they need; where everyone can access clean energy that doesn't contribute to climate change; where women and girls are afforded equal rights and equal opportunities (*38*)."

-United Nations Secretary-General, Ban Ki-moon

Many of the challenges outlined by the quote above are directly related to the chemistry we use and the chemical products we create in our society. Green chemistry is essential in minimizing impacts on humans and the environment, while also providing the many health, technological and other benefits that chemical products can afford. As the green chemistry field grows, we invite other institutions to contribute unique models for teaching green chemistry topics that can inspire even more departments and institutions to get involved. Current actions can be taken by faculty, departments, administrators, students, and other interested partners.

At this time the GCC is the only nationwide program specifically designed to encourage, empower, and celebrate entire departments of chemistry that transform their curriculum through green chemistry. The institutions that have begun this transformation have benefitted in many ways, some tangible and others that are intangible, by their participation in the GCC. Individual faculty who were lone proponents now possess a tool to encourage the others in their departments to follow their lead. Departments themselves now possess a process to encourage continual improvement in their asymptotic incorporation of green chemistry into all of the work in their department. Further, departments have a way of promoting themselves with prospective students as well improving their visibility with administrators who have previously found chemistry to be inac-cessible to the layperson.

And most importantly, the Green Chemistry Commitment is a pledge to our students that they will be trained in the most responsible way as they learn to control and describe the molecular species that will be inherently safer for themselves and their communities.

Joining the Green Chemistry Commitment is literally a transformative step that, given time, can lead to the ultimate goal of practioners in the field – the elimination of the adjective "green" from green chemistry. Indeed, one day, it is hoped, the chemical enterprise will look to the 12 Principles of Green Chemistry as the standard. GCC institutions will lead the way toward that ultimate goal of improved chemistry education for all of our students.

References

1. Millennials in Adulthood: Detached from Institutions, Networked with Friends. Pew Research Center, Social & Demographic Trends. http://www.pewsocialtrends.org/2014/03/07/millennials-in-adulthood/ (accessed June 20, 2015).
2. Musser, L. M.; Diamond, K. E. The Children's Attitudes Toward the Environment Scale for Preschool Children. *J. Environ. Educ.* **1999**, *30*, 23–30.
3. Hill, J. W.; McCreary , T. W.; Kolb, D. K. *Chemistry for Changing Times*, 13th ed.; Pearson Prentice Hall: Upper Saddle River, NJ, 2013.
4. Middlecamp, C. H.; Mury, M. T.; Anderson, K. L.; Bentley, A. K.; Cann, M. C.; Ellis, J. P.; Purvis-Roberts, K. L. *Chemistry in Context: Applying Chemistry to Society*, 8th ed.; McGraw-Hill Education: New York, 2015.
5. *Green Chemistry Education: Changing the Course of Chemistry*; Anastas, P. T., Levy, I. J., Parent, K. E., Eds.; ACS Symposium Series 1101; American Chemical Society: Washington, DC, 2009.
6. Ritter, S. K. Teaching Green. *Chem. Eng. News* **2012**, *90* (Oct 1), 64–65.
7. Green Chemicals Will Save Industry $65.5 Billion by 2020. Navigant Research. http://www.navigantresearch.com/newsroom/green-chemicals-will-save-industry-65-5-billion-by-2020 (accessed June 20, 2015).
8. Anastas, P. T.; Warner, J. C. *Green Chemistry: Theory and Practice*, Oxford University Press, 1998.

9. Anastas, N.; Warner, J. C. *Chem. Health Saf.* **2005**, *12* (2), 9–13.

10. Hanson, D. J. Gains continue for chemistry grads. *Chem. Eng. News* **2010**, *88* (Aug 23), 48–54.

11. Ph.D. Program in Green Chemistry. Boston Center for Green Chemistry, University of Massachusetts. http://www.umb.edu/greenchemistry/phd (accessed June 20, 2015).

12. Rouhi, M. Green Chemistry Earns a Ph.D. *Chem. Eng. News* **2002**, *80* (Apr 22), 42.

13. Greener Education Materials for Chemists. http://greenchem.uoregon.edu/gems.html (accessed June 20, 2015).

14. Chemistry Collaborations, Workshops & Communities of Scholars (cCWCS) Workshop. Green Chemistry. http://www.ccwcs.org/content/green-chemistry (accessed June 20, 2015).

15. Jackson, P. T. Case Study Approach to Green Chemistry Impacts on Science Facility Design and Operations: Regents Hall of Natural Sciences in St. Olaf College. In *Innovations and Renovations: Designing the Teaching Laboratory*; O'Connell, L., Ed.; ACS Symposium Series 1146; American Chemical Society: Washington, DC, 2013.

16. Case Studies for Organic Chemistry. Green Chemistry Commitment. http://www.greenchemistrycommitment.org/resources/curriculum-resources/case-studies/ (accessed June 20, 2015).

17. Vulpe, C.; Mulvihill, M. The Basics of Toxicology for Green Molecular Design. http://bcgc.berkeley.edu/toxicology-basics-green-molecular-design (accessed June 20, 2015).

18. Raynie, D. South Dakota State University, Brookings, SD. Personal communication, 2013.

19. CHEM342 – Mechanistic Toxicology was first offered in Fall 2011 and again in Spring 2014. Faculty are considering mandating this course as a requirement for the B.S. Chemistry degree.

20. CHE391/392 and 491/492 at Gordon College have four themes that rotate through a two-year cycle. One of the four themes is Toxicology for Chemists.

21. Toxicology Training for Green Chemistry Education Webinar, April 23, 2013, Michigan Green Chemistry YouTube channel. http://www.youtube.com/user/migreenchemistry (accessed June 20, 2015).

22. The Green Chemistry Commitment. http://www.greenchemistrycommitment.org/ (accessed June 20, 2015).

23. American College and University Presidents' Climate Commitment. http://www.presidentsclimatecommitment.org/ (accessed June 20, 2015).

24. Molecular Design Researh Network (MoDRN) is developing undergraduate curriculum. http://modrn.yale.edu/education (accessed June 20, 2015).

25. *State-of-the-Art Symposium. Toxicology and Environmental Impact in the Chemistry Curriculum: Science and Strategies for Educators.* 250th National Meeting of the American Chemical Society, Boston, MA, August 16–17, 2015; CHED program.

26. Green Chemistry at Gordon. http://www.gordon.edu/greenchemistry (accessed June 20, 2015).

27. Marasco, C. The Ivory Tower Goes Green. *Chem. Eng. News* **2008**, *86* (Sep 8), 64–66.

28. Green Chemistry Principle #2: Atom Economy. http://www.acs.org/content/acs/en/greenchemistry/what-is-green-chemistry/principles/gc-principle-of-the-month-2.html (accessed June 20, 2015).

29. E-Factor: Environmental Impact Factor. http://www.beyondbenign.org/K12education/msgc/17%20e%20factor.doc (accessed June 20, 2015).

30. Kay, R. D.; Levy, I. J. Student Motivated Endeavors Advancing Green Organic Literacy. In *Green Chemistry Education: Changing the Course of Chemistry* Anastas, P. T.; Levy, I. J.; Parent, K. E., Eds; ACS Symposium Series 1011; American Chemical Society: Washington, DC, 2009; pp 155–166.

31. Kwon, S. Y.; Levy, I. J.; Levy, M. R.; Sargent, D. V.; Tshudy, D. J.; Weaver, M. A. The Dose Makes the Poison: Measuring Ecotoxicity Using a Lettuce Seed Assay. In Greener Education Materials for Chemists (GEMs) Database. http://greenchem.uoregon.edu/gems.html (accessed June 20, 2015).

32. We have found the Acros Organic entries in Chemexper chemical directory to be particularly useful. See http://www.chemexper.com/ (accessed June 20, 2015).

33. Greener Solvent Alternatives. http://www.sigmaaldrich.com/insite_greener_solvent_alternatives (accessed June 20, 2015).

34. Boyd, J. Erskine College, Due West, SC. Personal communication, 2012.

35. 2014 Chemistry Rankings. U.S. News & World Report. http://grad-schools.usnews.rankingsandreviews.com/best-graduate-schools/top-science-schools/chemistry-rankings (accessed June 20, 2015).

36. Interdisciplinary Curricula. The Berkeley Center for Green Chemistry. http://bcgc.berkeley.edu/education (accessed June 20, 2015).

37. Graduate Ethics Course, Spring 2015. The Berkeley Center for Green Chemistry. http://bcgc.berkeley.edu/new-graduate-ethics-course-sp2015 (accessed June 20, 2015).

38. Ki-moon, B. Big Idea 2015: Sustainability Is Common Sense. https://www.linkedin.com/pulse/big-idea-2015-sustainability-ban-ki-moon (accessed June 20, 2015).

Editors' Biographies

Kimberlee Daus

Kimberlee Daus received her B.S. and Ph.D. in Chemistry from the University of Tennessee at Knoxville. Although she has served in a variety of roles since coming to Belmont University, including Assistant Director of General Education, Teaching Center Director, and Associate Dean, Kim's passion is teaching. Some of her favorite classes include *Better Eating through Chemistry* and *Your Brain on Drugs: The Chemistry of Drugs and Addiction.* She is an avid adopter of active-based learning and was the 2014 recipient of Belmont's teaching award. Her son, Pete, and daughter, Rebecca, continue to provide inspiration and support.

Rachel Rigsby

Rachel Rigsby received a B.A. in Chemistry from Kentucky Wesleyan College in 2000 and a Ph.D. in Chemistry from Vanderbilt University in 2005. She is currently an Associate Professor of Chemistry at Belmont University in Nashville, Tennessee. She currently serves as the University Learning Communities Course Coordinator and has taught a variety of chemistry Learning Community courses. She enjoys teaching organic and biochemistry and serving on the Executive Committee of the Tennessee Academy of Science as Managing Editor of the *Journal of the Tennessee Academy of Science.*

Indexes

Author Index

Subject Index

Printed in the USA/Agawam, MA
March 11, 2016

631876.001